Livestock Behaviour

Dedication

This book is dedicated to two Scotts.

Firstly, to the late T. Harry Scott, first Professor of Psychology at the University of Auckland under whose direction and encouragement while at Canterbury University R.K. became involved with the study of behaviour in a herd of dairy cows. Harry was killed in a climbing accident on Mt Cook in 1960.

Secondly, John Paul Scott, now Professor Emeritus at the Center for the Study of Social Behavior, Bowling Green State University, Ohio, who while at the Jackson Laboratories in 1958 took the time to encourage R.K. in further studies of farm animal behaviour.

Hamilton NZ 1983

Livestock Behaviour

a practical guide

Ronald Kilgour and Clive Dalton

Routledge
Taylor & Francis Group

LONDON AND NEW YORK

First published 1984 by Westview Press

Published 2018 by Routledge
52 Vanderbilt Avenue, New York, NY 10017
2 Park Square, Milton Park, Abingdon, Oxon OX14 4RN

*Routledge is an imprint of the Taylor & Francis Group, an informa
business*

British Library Cataloguing in Publication Data

Kilgour, Ronald
   Livestock behaviour.
   1. Domestic animals – Behaviour
   I. Title   II. Dalton, Clive
   636   SF 756.7

ISBN 0 246 11906 3

ISBN 13: 978-0-367-01983-9 (hbk)
ISBN 13: 978-0-367-16970-1 (pbk)

Contents

Foreword

By Professor Jack Albright

Why study behaviour?

Animal caretakers have been watching their herds and flocks for a long time so why all the current interest in animal behaviour? Until fairly recently, the science of animal behaviour has been restricted in academic circles to psychology. The relatively new science of 'ethology', which is the biological basis of animal behaviour in response to its environment, both animate and inanimate, has been accorded added stature by the award of the 1973 Nobel Prize in Medicine to three leading European ethologists working in this field – Konrad Lorenz, Nikolaas Tinbergen and Karl Von Frisch. Whilst Charles Darwin is recognised primarily for his development of the theory of evolution, he was also an astute observer of animal behaviour and could well merit the title of 'The Father of Animal Behaviour'. As the importance of animal behaviour as a factor in the management of the livestock enterprise is now being recognised, the study of 'ethology' or animal behaviour is becoming a recognised discipline within schools of agricultural sciences and veterinary medicine. This interest in animal behaviour is due in part to its implications in the economic production of livestock, in response to individuals and groups concerned about animal welfare and finally as a *bona fide* scientific endeavour.

Ron Kilgour and Clive Dalton have made great strides in cataloguing and clarifying various aspects of animal behaviour. They have shown that both environment and experience shape behaviour in animals. Behavioural observations often have been secondary to the main thrust of a research project, and frequently behavioural researchers have published in journals not well known to the animal scientist. It is important to have a clear understanding of an animal's behaviour under various environmental conditions for an intelligent analysis of results on nutrition, physiology, breeding and management. For example, a nutritionist who ignores behaviour may find that differences measured on alternative rations were confounded by the stress of isolation or crowding. A reproductive physiologist ignoring early rearing experience may be frustrated in evaluating individual breeding efficiencies of animals on trials. Breed, sex, age and nutrition are standard criteria for standaridising experimental trials. Geneticists must be aware of cross-fostering ('mis-mothering') in farm animals. The effects of animal behaviour now are recognised as important considerations in experimental and environmental design; without behaviour the whole animal is overlooked.

Careful observations needed

Animal behaviour is a 'whole animal' subject. Therefore, careful, analytical observations are necessary to solve problems. For example, when Charles Darwin visited a large cattle ranch in South America (1833) he observed one mob of some 10 000–15 000 head of cattle. Though they were thoroughly mixed at collection time and after stampeding, the cattle invariably came back into the same small and stable groups of 40–100 head which was their usual mode of existence. Although Darwin assumed that each animal of this large group could recognise all of the others, it is just as logical to believe that each animal knew only 40–100 other herdmates. It is interesting that world-wide, dairy farmers, for the most part, have settled on approximately these same numbers of 40–100 cows for their group size.

Doing an ethogram

Like Darwin, Kilgour and Dalton have presented methods of analysing and solving behavioural problems as well as doing behaviour inventories (ethograms). The authors have put forward to people and farm animals an 'unwritten contract' (welfare) between the two parties. In order to obtain beneficial results, there is a tendency in their thinking and writing to try and 'ask the animal'.

Species-specific behaviour, management recommendations and welfare concerns for cattle, sheep, goats, deer, horses, pigs, poultry and dogs have been covered in great detail. Cattle are further separated to cover both beef and dairy functions. The authors are to be commended for bringing forth the avoided areas of bulls and handler safety. Sheep and goats in this book have been separated, as they should be, due to their wide behavioural differences as well as in differing chromosome numbers. So called crosses between them described as geeps or shoats have been reported but most are non-viable. At the outset, the choice of deer for this text might seem unusual. However, New Zealand has met the growing market demand by providing farm-raised venison to the world.

Natural history of Selborne

Management and handling advice, training, likes and dislikes of the horse, and behavioural problems are covered in great detail for horses. It should be remembered that the stall is the birthplace of vices and gregarious horses get into trouble from being alone. Kilgour has stated that at least 4–5 sheep are needed to show normal behaviour and flocking instinct which is not present in single sheep. Likewise a single cow, horse, hen, etc. are all social animals (herd, band and flock, respectively) and in need of company. In Gilbert White's *Natural History of Selborne* (Letter XXIV on 15 August 1775) one is reminded of the importance of companionship:

> . . . 'For a very intelligent and observant person has assured me that, in the former part of his life, keeping but one horse, he happened also on a time to have but one solitary hen. These two incongruous animals spent much of their time together in a lonely orchard, where they saw no creature but each other. By degrees an apparent regard began to take place between these two sequestered individuals. The fowl would approach the quadruped with notes of complacency, rubbing herself gently against his legs; while the horse would look down with satisfaction, and move with the greatest caution and circumspection, lest he should trample on his diminutive companion. Thus, by mutual good offices, each seemed to console the vacant hours of the other.'

Kilgour and Dalton add to the concept that more is known about courtship, pheromones and behavioural signals between the boar and the gilt or sow than any other farm animal species. Also discussed were play, comfort behaviour, tail biting and cannibalism. Hens, which are the most confined of all the farm animals, are compared in the text from their modern status back to wild jungle fowl of South East Asia. The authors have covered farm dogs, of which more is known experimentally than other animals covered, concerning the effect of early experience, training, discipline and working concepts.

Applied behavioural terms

Behavioural terms from New Zealand and elsewhere appear in the text. Terminology such as 'easy-care' sheep, 'sheep-worrying' dogs, hind (female red deer), SAG (sexually active groups) and others appear and add to the language of animal behaviour and animal welfare.

Of the observations and recommendations put forth in this book many, *but not all*, will be known by livestock producers, advisers, technicians, academics, and students of agriculture and veterinary medicine. The authors raised and expanded upon the long neglected issues of transport and slaughter. Although a great amount of information was covered by Kilgour and Dalton, more research work in the area of animal behaviour and care is needed. This thoughtful book will be an important reference for those persons interested in farm animal behaviour as related to the animal's welfare.

Preface

The aim of this book is to bring together the essential information on animal behaviour for those concerned with the husbandry, management and welfare of farm animals.

Farm animal behaviour, or applied ethology, is a rapidly developing science but the information is scattered in many different places. Most text books on behaviour cover principles rather than practice, and relate to species that have little value in present agriculture.

This book is intended for diploma and degree students in agriculture and veterinary science where animal behaviour is part of their courses on animal production and health. It also aims to help students of general biology and applied psychology who are seeking information on the practical importance of ethology. Technicians and advisers concerned with problems of livestock management and people concerned with animal handling facilities, abattoirs and housing will also find the book useful.

The book provides information to make fuller use of labour, limit handling problems, reduce accidents to both men and beasts, and generally increase the wellbeing and productivity of farm livestock. It will be of interest to those concerned with animal welfare as an information source on current farming practices.

Chapter 1 covers definitions of behaviour and some of the principles which relate to domestication, man's relationship with livestock and the implications for husbandry which this produces.

The behaviour of cattle, sheep, goat, deer, horse, pig, hen and dog is described in chapters 2—9. Within each chapter, birth and rearing, early life up to puberty, sexual development, mating and reproduction are covered. Within each species, special behavioural topics such as milking, intensive housing, animal training techniques and vices are also considered.

Chapter 10 draws out some of the general principles which thread through the species discussions and briefly introduces the reader to specialised fields such as behaviour during transport and slaughter, and general animal handling.

Throughout the text, documented information has been drawn on and where available key references have been provided at the end of each chapter for those who want to study further. We have also drawn widely on farmer experience both verbal and where reported in the farming press. The practical advice given is thus based on proven research and the experience of the authors and many practical farmers around the world.

Six appendixes and a glossary have been added to provide a summary of useful information and further reading.

Acknowledgements

Thanks are due to many people who have helped in the production of this book; to the publishers for seeing the need for such a text and their advisers for suggestions about the manuscript; to Professor Jack Albright for the foreword; to other authors for permission to use their material and their ideas. We have drawn extensively on the experience and expertise of friends and colleagues from a wide source. They have all given valued comment and suggestions. They are:

General: Dr W.G. Whittlestone; Mr J.D. Kilgour; Miss K.J. Bremner.
Cattle: Mr K. Macdonald.
Sheep: Mr A.E. Uljee; Mr J. Dobbie.
Goat: Mr G. Batten; Miss J. Radcliffe; Mr J. Woodward.
Deer: Dr J.L. Adam; Dr G.W. Asher; Dr G. Moore.
Horse: Mr H.F. Dewes; Dr G. Waring.
Pig: Dr J.L. Adam; Mr K. Hargreaves.
Hen: Mr P. Josland; Mr N. Humm.
Dog: Mr I.R. McMillan; Mr and Mrs A.R. Young; Mr W.H. Bishop.

Typing of the drafts was done by the typing pool at the Ruakura Agricultural Research Centre. We are very grateful for their skilled assistance and willing co-operation.

Artwork was done by Ms Pauline Hunt of the Ruakura Research Centre, and Mr D.J. Walmsley of the Ministry of Agriculture and Fisheries Display Centre, Hamilton.

The figures were redrawn from various sources and these are acknowledged in the figure captions.

Acknowledgement is also due to our employer, the New Zealand Ministry of Agriculture and Fisheries, for the opportunity and experience gained in our current positions which have enabled us to write the book.

For a book which highlights the importance of social contact, we owe very special gratitude to our wives and families for accepting comparative isolation during its writing.

1 Introduction

What is animal behaviour?

'Behaviour' is a term used widely in many sciences but in this book it covers, 'the patterns of action observed in animals which occur either voluntarily or involuntarily'. An animal's behaviour provides information on a wide range of factors such as breathing, eating, drinking, fighting, mating and milking. Observations of external behaviour can often lead to deductions about the internal state of the animal.

The term Ethology is now used to describe the science of animal behaviour although others would define it as the 'biology of behaviour'. More recently farm animal behaviour studies have come to be termed 'applied ethology', which covers each farm animal species and such topics as stockmanship, stress, transport and welfare.

Differences between species

Even the casual observer of farm animals is aware of the wide differences in behaviour between species, and these have a profound effect on farming systems. Species-specific responses which are carried out in the same fixed way are not altered by learning. Other species behavioural responses are more flexible and can be changed by learning.

Farmers need to work with, rather than against, the animals' known species-specific behaviour patterns and in the short term it is easier to fit husbandry to the animals than change the animals by selection to fit what may in fact be inappropriately designed units.

The differences between farm animal species is further highlighted in the sensitivity and the way they use their eyes, ears, noses and other sense organs. This can also differ greatly from man's sensory abilities. The challenge to the stockman is to appreciate what each species can see, smell, hear and feel in order to anticipate their reactions during handling. This is the art of stockmanship.

Domestication

As man has only domesticated about sixteen of the 3000 animal species, it cannot be concluded that all worthwhile domestication has been completed. Domestication is a continuing process and the consideration of behaviour is fundamental in any attempts to extend domestication to new species.

Confusion arises over definitions of terms such as wild, feral, tame and domestic. A wild animal has been left alone in its habitat and not exposed, harassed or hunted by man. Few animals today would fit this definition, so it is more realistic to say that wild animals have, in general, been left in peace with minimal disturbance from outside.

A feral animal is one that was formerly domesticated but has been released or has escaped and returned to a semi-wild state. There are many examples around the world of feral cats, dogs, sheep, cattle, pigs, horses, goats, donkeys and camels. These feral populations can provide valuable behavioural information which can be applied to the domestic counterpart.

Taming is not domestication but describes the process of reducing the flight distance of the animal when approached. Tame animals may let you walk up to them and even touch them without being restrained.

Domestication is dealt with more fully in other texts where there is a range of definitions (see references 2, 5, 6, 10, 11, 12) for those wanting further information. Hale[1] provides a list of the advantageous traits in certain species which allow them to live more effectively with man and allow man to manipulate them to mutual advantage. This approach has been developed further by Kilgour[3,4] and is described as the 'domestic contract'.

Man's domestic contract with animals

In this view, man's relationship with farm animals is thought of in terms of an unwritten 'contract' between the two parties — it is a trade-off. For the economic benefits man obtains from livestock they must come to accept a number of changes:

- Control of breeding — the number of sexual partners is reduced or artificial insemination is used.
- Changed survival patterns — the weak are helped to survive and disease and parasites are controlled.

- Altered nutrition — feed quantity and quality is manipulated and the range of foods on offer reduced.
- Genetic selection can lead to changes which in time may set the domestic animal apart from its wild counterpart.
- Reduction in freedom and choices.

To honour this contract it behoves man to know and have accurate information about the animal's physiological and behavioural needs. This is especially important when farming systems are altered and become more intensive and restrictive. The study of farm animal behaviour plays an important role within this contract and helps the husbandman assess and meet the animal's welfare needs. The contract should be symbiotic where both partners benefit. It should also be regularly re-examined as technology and economic pressures act to force change. McBride[7] has presented a useful framework to show how man's contract with farm animals differs from the interactions taking place for wild animals.

Welfare

The welfare of farm animals only takes on proper meaning within the terms of the domestic contract. Otherwise problems arise with the definition of what constitutes welfare. Some common consensus must be reached by all who keep animals for research, sport, entertainment and pleasure as well as farmers, so that man's responsibilities to these animals can be examined and re-examined. For some, welfare has become a very emotive word and it has even been used as a substitute for stress.[8] A range of definitions is presented for interested students in appendix 1. The concern over what constitutes 'intensive' and 'factory' farming is also part of this debate and definitions of these from the literature are given in appendix 2.

Concern for animal welfare issues has tended to grow with urbanisation. Few people in cities have experience of rural life and the gap between 'town' and 'country' widens. Both the media and legislature gather increasing support from urban centres, and the urban views which are propounded will have increasing impact in the future.

Viewpoints on welfare

There are five broad viewpoints on welfare.[3] These are:

- Welfarists — people mildly or strongly opposed to many or all forms of intensive farming. They may include those who believe animals should not be exploited for any reason.

- Farmers or operators – who have an interest in, or gain a livelihood from, farming animals.
- The public – who tend to be disinterested unless their attention is focussed on an issue by the media, and who would generally resist paying extra to cover the costs of food produced under improved-welfare systems.
- Research workers – whose job is to provide insights into the response patterns and behavioural needs, and who build up the ethogram or inventories of farm stock.
- The animal – whose needs should be accurately defined by good research work both on and off the farm.

Further discussion points on each viewpoint are given in appendix 3.

Welfare in practice
Farmers need to ask themselves practical questions such as:

- Are the animals producing normally?
- Are they healthy and free from injury?
- Is their behaviour normal?
- Are the animals handled or housed without undue stress?
- Are they handled in accord with their species-specific behaviour patterns?
- If the system is adversely affecting them, can it be modified and made more acceptable?
- If the systems should be abandoned what are the appropriate and viable alternatives?

The motivation for good farm animal welfare grows from the concerned and informed farmer's response. Legislation will do little to change human behaviour or affect human motives. Laws are needed to cover cases of gross cruelty, but codes of practice are more helpful guidelines to improve and suggest to farmers ways in which the welfare of livestock can be improved. The husbandman must finally be responsible for the animals in the system and their management, as the terms of the domestic contract are upheld. Codes can act as a guide.

Analysing behaviour problems

When problems arise which have or are suspected to have a behaviour component, a number of questions need to be answered. These are:

- What is the animal doing – how is it behaving?
- Would this behaviour be expected at this time and place?
- Is the behaviour normal or abnormal (anomalous)?
- Why is the animal acting in this way?

Each question follows on from the previous one and the answers should be obtained by referring to the species ethogram or the catalogue of precise detailed responses which make up an animal's behaviour. Examples of ethograms for sheep and goats[1] and for dogs, wolves and foxes[9] (see appendix 5) are available, though a little sketchy in places. The system of behaviour classification used as the basis for the sheep ethogram can be a useful guide when building up ethograms for other farm species. Where ethograms are not yet available, observations and detailed recording will need to be made by the stockman or researcher. One important practical problem and the way it was approached to solve it, is given in appendix 4.

Social behaviour

The domestic contract involves highly social farm animals. These animals cannot be successfully managed if their social behaviour is ignored. Social behaviour is defined as 'the regular and predictable behaviour that occurs between two or more individual animals'. Although farm animals are usually run in groups, they are rarely natural groups but are arranged by farmers according to sex, size, age or breed and this can lead to additional pressures or harassment of one animal by another.

Within groups of animals various social structures exist. During movement, a regular movement order can be seen in herds or flocks, and a social dominance hierarchy or 'peck' or 'bunt order' can be set up in groups which operate together for any length of time. The social dominance order affects many aspects of farm animal management particularly when animals have restricted access to space, mates, food or water.

Animals not only interact with their own species, but they respond to other species' members. They may be treated as 'neutral' and not posing a threat, or as predators. The dog is often used as a domestic animal to work sheep or cattle and they react to the dog as a predator. The domestic contract set up by man also involves a social interaction where man may be treated as a social equal or more dominant partner (man/dog); neutral (man/cat) or as a threat (man/sheep; man/cow; man/horse). The success of these social interactions may also depend on whether the animal has had some positive or negative early social experience with man. If the larger farm species have been mishandled and have come to associate handling with pain and fear, they may remain a potential danger to man.

In this chapter the nature of the unwritten domestic contract between man and farm livestock has been outlined and some of the implications indicated. The welfare debate has been set in its proper context. It is now appropriate to turn to a consideration of the behaviour· important for practical management for each species in much greater detail.

References

1. Hafez, E.S.E., Cairns, R.B., Hulet, C.V. and Scott, J.P. (1969) 'The behaviour of sheep and goats', Chapter 10 in *The Behaviour of Domestic Animals*, 2nd edn, ed. E.S.E. Hafez, London: Bailliere, Tindall and Cassell, pp. 296--348.
2. Hale, E.B. (1969) 'Domestication and the evolution of behaviour'. In *The Behaviour of Domestic Animals*, 2nd edn, ed. E.S.E. Hafez, London: Bailliere, Tindall and Cassell, pp. 22–42.
3. Kilgour, R. (1980) 'Animal welfare – the conflicting viewpoints'. In *Behaviour in Relation to Reproduction, Management and Welfare of Farm Animals*, eds. M. Wodzicka-Tomaszewska, T.N. Edey and J.J. Lynch, Armidale, NSW: University of New England Press, pp. 175–82.
4. Kilgour, R. (1983) 'Stress and behaviour: An operational approach to animal welfare'. *Proc. C.E.C. Sci. Sympos. on Animal Housing and Welfare*, Aberdeen, July 27–29, 1982, pp. 1–13.
5. King, J.M. and Heath, B.R. (1975) 'Game domestication for animal production in Africa'. *Wld. Anim. Rev.* **16**, 23–30.
6. Lorenz, K. (1953) *Man Meets Dog*. Harmondsworth: Penguin, pp. 1–198.
7. McBride, G. (1980) 'Adaptation and welfare at the man–animal interface'. In *Behaviour in Relation to Reproduction, Management and Welfare of Farm Animals, Rev. Rural Sci.* IV. Univ. New England, N.S.W. pp. 195–198.
8. Murphy, L.B. (1978) 'A review of animal welfare and intensive animal production'. *Q. Dept Primary Indust., Poult. Section. Publ.* pp. 1–101.
9. Scott, J.P. and Fuller, J.L. (1965) *Genetics and Social Behavior of the Dog*. Chicago: University of Chicago Press, pp. 468.
10. Spurway, H. (1955) 'The causes of domestication: an attempt to integrate some ideas of Konrad Lorenz with evolution theory'. *J. Genet.* **53**, 325–362.
11. Thines, G. (1978) 'Ethology and animal production'. *Proc. 1st Wld Congr. Ethol. appl. Zootech.* Madrid, E-OP-1, pp. 13–20.
12. Zeuner, F.E. (1963) *A History of Domesticated Animals*. New York: Harper and Row, p. 1.

2 Cattle

Cattle have played an important part in man's development since prehistoric times and have provided meat, milk, clothing, utensils and draught. They are also used as currency, to denote wealth and can have a religious significance.

Within less than 100 years man has taken the calf from the cow and turned it into an efficient machine to turn plant products into milk and reproduce regularly to achieve this.

Behavioural definition

Cows are large, hairy ruminants living in herds which roam over a large area of grassland. The female withdraws from the herd to produce one (rarely two) precocial young which soon stand and suckle 4 to 6 times a day. They 'lie-out' away from the dam for at least the first week of life during the daytime. These lying-out patterns together with the head threat act to set greater distances between cows in the herd than sheep in a flock. In small herds, strait-line social dominance ranks are found and these are stable over many months.

As a ruminant the cow has 3 major grazing phases per 24 hours, totalling between 7 and 9 hours interspersed with resting and rumination periods. Heaviest grazing is from near dawn, before and into dusk and about midnight.

Although cattle mate at any season of the year they maintain a seasonal pattern with a gestation of 9 months, and the cow rarely cycles during the early high-suckling period. The young stay in the cow herd after weaning but the males leave this group at about 2 years of age, join the male sub-group and fight their way to the point where they are able to mate some of the cows or even become 'king' bull. As the herd grazes widely over the range, a mating strategy evolved whereby cows in heat come together and begin riding one another. The bull can determine by seeing distant mounting, that some females are in oestrus and move to this part of the range for mating.

Birth

Behaviour before birth

Approaching calving, the cow moves from the centre of the herd towards the outside of the group and usually becomes much more submissive, avoiding conflicts even with animals lower down the social order. She normally attempts to find a quiet, sheltered place to give birth away from other cows and human disturbance. Disturbance of the cow once she has chosen a birth site will delay calving for periods ranging from minutes to even hours. Dufty[9] showed how birth problems increased as the inspection and handling of the animals at calving increased. This is shown in table 1.

It is good practice to reduce disturbance of the calving cow where possible. However, in some cases disturbance can be exploited by the stockman when calving cows indoors. Some use the technique of feeding indoor-calving cows between 10 and 11 p.m. so that calving will be delayed until early the next morning allowing the stockman a night of undisturbed rest.

Behaviour during birth

There are few reliable signs of the approach of birth in the cow. Among those used are restlessness, bagging-up of milk in the udder, relaxation of the pelvic ligaments around the tail, swelling of the vulva and exudation of mucus from the vagina.

Labour is usually described as being in three stages. The first stage is completed when due to contractions of the uterus, the membranes and fluids which signal the birth are pushed out and frequently break (breaking the waters).

The second stage is when uterine contractions increase to expel

Table 1 The effects of increasing disturbance on calving difficulties[9]

	% stillbirths	% dystocia (difficult births)
Left quietly at pasture	8	14
In yards – not observed	0	19
In pens – not observed	0	50
In pens – observed	0	65
Regularly handled	27	0
Group housed indoors	35	0

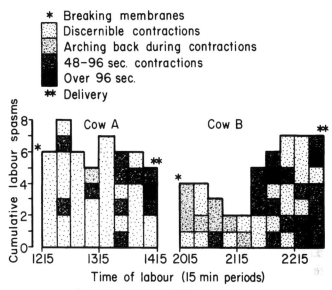

Figure 1 Partograms (records of labour) for two cows. Each column records spasms every 15 minutes[9]

the calf. Normal presentation is head and both forelegs first. The third stage is when the placenta (afterbirth) is expelled. Normally at this stage the cow will be licking the calf and most cows will chew or eat the afterbirth.

The patterns of labour can be recorded in a partogram (see fig. 1).

In the cow, the contractions are usually well-spaced and there are about 4 to 7 major contractions each 15 minutes. Working in 15 minute intervals, a square of the partogram is hatched in for every contraction when they occur and after 15 minutes, a new column is used. From a number of these partograms it has been found that normal labour in a cow is about 2 hours before the calf is born.

Cows normally get to their feet after the severe contractions which push the calf's head and shoulders through the birth canal. When this happens the calf drops to the ground and the cow turns to lick it. A vigorous calf will shake and free its head of any membrane and start breathing.

In prolonged and difficult births (called dystocia) or where the cow is ill (e.g. hypocalcaemia), she may be too weak or exhausted to get up and lick the calf. The calf may also be exhausted and may suffocate in the membranes unless assisted. In difficult calvings where considerable traction on the calf is needed to deliver it, the cow is often injured (e.g. torn uterus or temporary paralysis). Such cows take little interest in their calves.

Bonding of cow and calf

Only a few minutes are needed after birth for the cow—calf bond to be established[18]. From then on the cow will bunt or kick alien calves. Thus mothering-on extra calves requires a few tricks to confound the cow. It is often claimed that dairy cows settle down to milking quicker if calves are removed at birth rather than at 2—4 days when the calf is old enough for sale or slaughter and the flow of colostrum has finished.

Cow's reactions to the calf

The cow normally licks her newborn calf and vigorous licking may make the unsteady calf fall back to the ground. An experienced mother stands still while the calf moves forward past the cow's head towards the udder. The cow then licks the tail and perianal area of the calf which is parallel to the cow seeking the udder (fig. 2).

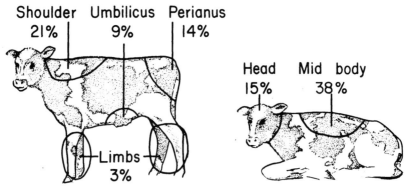

Figure 2 Body areas on calf when standing and lying licked by the cow immediately after birth (after Aitken, V.R. (1978) *Min. Ag. Sci thesis*, Massey Univ., N.Z.)

Some first calvers may show alarm as the calf approaches them and back away. When the calf approaches from the side she may swivel to face it and again prevent the calf moving down the flank to find the teats. Some cows may bellow and attack their own calves and this behaviour is sometimes aggravated by the presence of men and dogs.

Stress to some cows calving in outdoor herds is heightened by other cows coming to sniff and inspect the calf especially if the investigating cow is more dominant. It is unusual however for cows to steal calves from other cows.

It appears that licking and cleaning the calf helps the cow to recognise her own calf and thus refuse strangers. It also removes sources of smell and afterbirth which lessens the risk of attracting predators (e.g. dingos in Australia).

Eating the afterbirth is a normal behaviour pattern in most hooved species whose young 'lie out' in undergrowth for long periods while their dams graze. Some stockmen prevent the cow eating the afterbirth in case she chokes while others consider it a rich source of hormones and protein.

Calf's reactions to the cow

After becoming steady on its legs, the calf approaches the cow with its head at a set height searching for the teats which it locates by trial and error. It pushes under the dam's front legs as frequently as the back legs (see fig. 3).

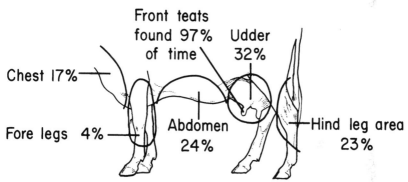

Figure 3 Parts of the cow nuzzled by the calf when first seeking the udder after birth (after Aitken, V.R. (1978) *Min. Ag. Sci thesis*, Massey Univ., N.Z.)

The licking and sniffing of the calf's perianal area by the cow assists the calf, particularly if the cow stands still. The front teats are generally found first. In cows with abnormally large or pendulous udders, teat searching by the calf will take longer. In dairy cows with long pendulous udders the calf needed about 2 hours longer to locate the teats[43].

Most calves have their first good feed within 5–6 hours after birth and will suckle about every 5 hours during the first day, changing to suckling intervals of 3 hours during the next few days. The calf suckles from both sides of the cow and occasionally from the rear until a definite pattern is established. In these early days of life about 10–15 minutes suckling is sufficient to satisfy the calf's appetite. With more than one calf suckling, there is considerable social facilitation as one calf encourages the other to suckle. Calves soon learn to bunt the cow's udder possibly to stimulate a greater milk flow, and as they grow this may hurt the cow and elicit a kick. Experienced calves stand close in to the cow during suckling to avoid being kicked.

Calves rest for about 18–20 hours out of the 24 during their early lives and sleep for considerable periods. Rest is very important for satisfactory early growth and development.

Rearing

The cow–calf daily rhythm
In herds at pasture, observations show that at dawn, cows and calves are lying together in groups where the herd has camped overnight. When the cows get up, the calves have their first suckle then turn away from their dams and lie down.

The dams then start to graze leaving the calves in a group or 'creche'. The nearest cows are the first to return to the calves when a calf gives an alarm bawl. They do not guard the whole group and are only concerned with their own calves.

About mid-morning one or two calves will call and their dams will return and this will start a mass cow–calf suckling session through social facilitation. This same pattern is repeated in late afternoon and in the evening the cows return to their calves for a feeding and grooming session. As the light fades they form up with their calves in the night camp.

This typical lying-out behaviour where the calves are left in groups makes precise identification of calves to their dams difficult (in contrast to ewes and lambs). After 2–3 weeks the calves are more closely associated with their dams but are then more active and difficult to catch for tagging.

Artificial rearing
The natural daily rhythm and the infrequent suckling makes artificial rearing much easier than in frequent-suckling species like the horse. Once or twice a day feeding systems can thus be devised without stressing the calves.

There are many systems used to rear calves artificially (i.e. not suckling their own dam). Some of these are:

- *Ad libitum* feeding of warm milk from automated feeders.
- *Ad libitum* feeding of cold milk (colostrum, whole milk or reconstituted powdered milk) from bulk feeders.

- Restricted feeding using buckets.
- Restricted feeding using rubber teats.

There are variations on these basic systems in different countries. Where labour is restricted, fully automated, 24-hour access *ad libitum* systems are common. However it was shown[13] that even in these systems calves still did not feed very frequently. Twins had similar feeding behaviour patterns.

Teaching calves to suckle from rubber teats at the same level as a cow's udder is much easier than teaching them to drink from buckets. In the latter the calf has to drink head down with the nostrils near the level of the milk. This has to be taught by encouraging the calf to suck the fingers and slowly pressing its head into the milk. With experience of sucking, the fingers can be withdrawn as the calf drinks.

In teaching calves to adapt to any system the following principles of learning can be useful:

- Build on the calf's natural suckling response.
- Keep the feeding system simple and natural, e.g. the teats at cow udder height.
- Teach the routine in simple stages. Do not rush the task and do not punish the animal.
- Quietness and patience are valuable.
- Establish and maintain a routine.

A summary[38] of humane considerations in calf rearing stresses the importance of keeping calves in groups to ensure appropriate resting behaviour, social and activity behaviour. Milk should be offered at least 3 times a day and ingestion should be slowed down with a valve to take 10 minutes to consume. Solid fodder should be offered at an early age and the calf's surroundings should provide plenty of stimuli to allow exploration and play.

Social order in calves

Many workers have observed the early development of a social order in a group of calves. This has been seen especially in automated systems with continual access to a limited number of feeding positions (teats). Size and age seem to be the main factor in establishing the dominance order regardless of the feeding system. Social orders established in groups of dairy calves have been observed to last until the animals entered the herd.

In some breeds, especially of *Bos indicus*, social facilitation at feeding means that enough feeding points should be provided so that the group can feed together. Otherwise the fed calves will move away followed by others that have not suckled.

Research[5,6,48] has clearly shown that group rearing produces calves that are better adjusted than calves reared in isolation. Group calves are more vocal and compete better in social hierarchies, especially where food supplies are limited.

Calves have also been shown to learn quickly to discriminate between objects (black versus white and large versus small). It has been suggested that cattle could be selected for learning ability.

Sucking problems

Intersucking or the habit of sucking each other is common among dairy calves. After finishing their milk, especially if it is taken quickly, calves often continue to suck each other's ears, navels, teats or parts of their pen. Solutions to the problem include:

- Make the calves work harder and longer for their feed so that the desire to suckle declines soon after they are satisfied.
- Tie up calves after feeding or separate them.
- Apply muzzles.
- Put a repellent such as Stockholm tar or dung on ears, teats and navels.
- Give them access to dry feed.

The best solution to the problem is the first point above. When the suckling reflex in the calf is triggered, it lasts for a set period of time, even if all the milk has been drunk.

Practical experience shows that separation after feeding is effective but not very practical and muzzles and repellents involve extra work. Suckling of appendages needs early recognition otherwise it quickly spreads in groups and will continue into adulthood. Offering dry food is useful as is delaying the grouping of calves after 4 weeks of age.

Fostering on nurse cows

Cows vary greatly in their willingness to accept alien fostered calves. Normally after the cow has bonded to its own calf within minutes of birth, it cannot rebond on to another alien calf. However, many systems require fostering of calves and some methods used to achieve this are described below:

- Remove the cow's own calf before she has licked and smelled it and introduce the alien calves as soon as possible to the cow after they have been rubbed with birth fluids (collected in bucket). This is rarely possible in practice.
- Remove the cow's own calf after 2–3 days and pen her with some alien hungry calves.
- Blindfolding the cow while alien calves are introduced to her.
- Use odours spread over the calf to confuse the cow e.g. neatsfoot oil, perfume.
- Restrain the cow tightly in a stall or race before introducing the foster calves[10].
- While leaving her with her own calf, introduce an alien calf and teach it to suckle through her back legs[21].
- Tie an alien calf to the cow's own calf with a pair of dog collars and 40 cm of chain which includes a swivel. Cows come to accept the alien calf more quickly as both the calves must suckle at the same time.
- In a single-suckled beef herd the dead calf's skin may be placed over the foster calf until it is accepted by the cow.
- Put cow and calves in a small stall and give the cow a vaginal douche using a solution of 5 ml of veterinary iodine in 250 ml of water. This is sufficient to do 3 cows. The vaginal irritation and subsequent straining it induces often encourages the cow to accept the calves. This is not a recommended method.

Successful calf rearers ensure that groups of calves fostered on to cows are all the same size. It does not seem to matter if they are of a different colour. They are also very selective in the cows they keep as foster dams as some animals are more amenable and accept any calf without problems, viz. 'permissive mothers'.

Calf housing

Calf housing is important in colder climates and in intensive production systems where large numbers of calves are reared. Requirements are reviewed in references 37, 49. Disease can build up in the calf house especially where calves have been purchased and moved over from many different farms. Ventilation and temperature are critical factors in calf comfort and health[7].

Most behaviour and welfare attention has been given to calves held in individual, small and narrow pens or crates where they cannot turn round or see their neighbours. This system was used for white veal production where low-iron all-milk diets were fed in reduced lighting environments. It has been demonstrated how-ever that acceptable veal can be produced by calves which are group-fed, loose-housed on balanced diets. and that group-fed calves grow faster than those reared individually[27].

Anomalous or 'abnormal' behaviour was often seen in calves kept in crates such as excessive chewing of the crate, licking the walls as well as licking themselves and other calves. This latter results in the build-up of hair balls in the stomach. It has been pointed out that coat licking has no deleterious effect and that the hair balls produced may act as a physical replacement for rough-age. One study[25] showed how calves soon learned to reach around the end of their cubicle to lick each other and could lick up spilt milk from the dung channel which caused infections. Most concern has been over pen size and the types of flooring in the pen to provide comfort as bedding was not allowed in the interests of consumer demands for white flesh. Education of consumers is still called for.

Simple shelter-type housing is usually provided for groups of calves reared at pasture. The calves may graze freely and use the shelters communally or they may be tethered to individual shelters fitted with water or milk nipple feeders and moved regularly on to fresh pasture. Behaviour problems rarely arise in such systems.

Treatment of surplus calves

In many dairy areas, surplus calves may go for slaughter at a few days old as 'bobby' calves for veal, or they may be transported to other areas for rearing. Studies that have examined calf behaviour during transport show that exhaustion is common and young calves need special care to avoid dehydration and digestive upsets. Using plasma corticoid levels as measures of stress in calves it was shown that transport, regardless of age, was significant in causing stress while castration and dehorning were less stressful.

The minimum welfare and hygiene recommended standards for the care, handling and transport of 'bobby' calves in New Zealand, are as follows:

- The calf should receive colostrum soon after birth and be ad-equately fed.
- Outdoor calf paddocks should be well-drained, free of hazards, sunny and sheltered.
- Indoor housing and pens should be designed for easy cleaning

and be safe (i.e. no projections or poor flooring). There should be a dry area for resting.

- Overcrowding should be avoided and all feeding utensils cleaned and disinfected after each use.
- Before collection for sale or slaughter calves should be at least 4 days old, have dry navels, worn and flat hooves, and be robust and alert.
- Collection pens should have slatted floors, be in sheltered positions, have adequate room for each calf and be thoroughly cleaned after each day's use.
- Calves should be transported in clean vehicles with protection against wind, excess heat and cold with doorways and projections covered to prevent injury.
- Overcrowding and physical injury should be prevented and calves should be handled by being lifted bodily or pushed along gently. Calves should not be thrown, dragged or pulled by the head.
- Drivers should drive to avoid injury and stop every 2 hours to check the calves. Calves should not travel continuously for more than 20 hours without rest.

Studies of calf behaviour on long journeys[23] concluded that:

- As animals settle in to a behaviour pattern during the first 2 days of travel, considerable care should be given.
- Feed dry fodder to the calves before transit. It will help keep the dung hard and dry.
- Pens in trucks should hold 10–12 animals and there should be adequate head room. If head room is limited, they lie down and get trampled.
- In larger groups animals can rock together and keep away from the pen sides.
- There are no set positions for travel and constant adjustments are made as conditions change.
- Stock should be loaded after the night's rest and early morning feed. After 25 hours travel, calves took exercise, food and water in that order. Animals lose condition during transport and need good feed and rest on arrival at their final destination.

Slaughter

Calves become insensible about 85 seconds after their throats have been cut so any method used to produce unconsciousness till death must last this long. Electric stunning of the head only produces insensibility for about 30–40 seconds and cannot be considered humane[2]. A head-to-back stunner does upset heart beat and prolongs unconsciousness for the required time.

Adult bulls penned with strangers after arrival for slaughter tend to fight and their stress levels can be raised to the point where muscle pH is high and the meat is dry and dark-cutting. If bulls remain quiet and in their own groups the problem generally does not arise. Some type of percussion stunning before slaughter is the normal practice except in Jewish ritual slaughter where special holding and tipping devices are used.

Reproduction

The male

In herds of wild cattle, cows and calves tend to stay together in one herd for almost 2 years. After about 18 months old the males dominate their dams and soon after leave this herd and join the all-male group. From then on, they have to fight their way up the social hierarchy until they reach the 'king bull' status and gain mating rights[41].

With few exceptions, most threatening, mounting and fighting occurs after 2 years of age. Calves ride each other in play from an early age but it is not until puberty (yearling stage) that mounting takes a more functional connotation. By 4 years of age, bulls become very territorial and at pasture space themselves out. Although they regularly engage in threat and head sparring in various fighting holes dug in the pasture, they return to and remain in their own territory.

The patterns of threat, fight and flight have been documented, albeit rather sketchily, in the literature[41].

The array of head and body positions and their meanings as presently understood are shown in fig. 4.

The direct threat is head-on with head down and shoulders bunched. If the opponent such as another bull (or man) withdraws to about 6 m, the pressure will drop and both bulls will turn away. If not, the bulls may circle, drop into the cinch body position, or start with head-to-head or head-to-body blows.

Male riding problems

When bulls run in groups, such as for fattening or awaiting progeny results, problems arise as a result of excessive riding or mounting. Bulls ride each other head on, or over the rump and this activity seems to vary in intensity over the day. It is during this riding activity

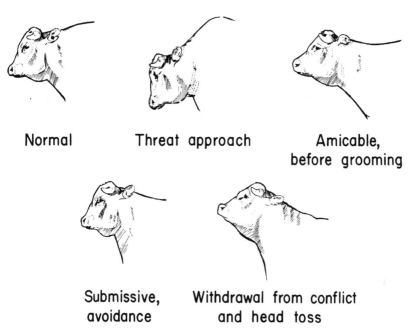

Normal   Threat approach   Amicable,
before grooming

Submissive,   Withdrawal from conflict
avoidance   and head toss

Figure 4 Head positions of bulls and their meaning for handlers[41]

which may occur in groups that most injuries, such as damaged penises and broken legs, may occur. Bulls seem to pick on a submissive member of the group and ride it to point of exhaustion after which they will pursue the next most-willing animal. The only solution is to remove the animal being ridden and the instigators of the riding outbreak if they can be located. The bulls riding others and those being ridden keep changing.

Suggestions to reduce riding and fighting, and hence injuries are:

- Avoid undue disturbance of the bulls.
- Graze them at a low stocking rate hence avoiding too frequent shifting.
- Allow older bulls room to establish territories or keep their space.
- Some drugs may be used (check local regulations).
- Some farmers claim that putting an old bull, a horse, a horned billy goat and even a roll of netting in the paddock will divert their attention and reduce problems.
- Adopt any procedure which will reduce the time left idle.

It is usually difficult (often impossible) to return an injured bull to the group, as it will be ridden again. If bulls have to be re-introduced to groups some suggestions are:

- Spray-dip all the bulls in the group including the stranger, and immediately return them to pasture.
- Make sure that on entering the new field or grazing area that all the bulls have room to scatter and graze at least 6 m apart.
- Starve them for a day, shift them to a new area at dusk and re-introduce the strange bull at this time.

Male sexual drive (libido)

Bulls vary in their willingness to mate females in oestrus and a test has been devised to assess their libido[3]. This is done by recording how many times groups of bulls (2–3) mount non-oestrous females or even other males secured in head bails. The bulls being tested are allowed a set period (e.g. 15 minutes). Other bulls waiting are kept close by to see the activity and be stimulated. The animals used as mounts should be changed regularly to avoid injury and exhaustion.

Blockey[3] showed how performance in this libido test was an indicator of paddock mating performance. However, care is needed in setting up the test to avoid biased results. The first bulls to be tested in the day are inadequately stimulated and should be re-tested later. Tethering the stimulus animals may be difficult and stressful for the animals involved. Any disturbing factors (e.g. dogs) should be removed from the scene.

High libido is only needed in bulls that have to serve many cows in a short period, or that have large areas over which to range. Libido and fertility may not be related.

Buffalo research[8] showed that two bulls were needed for satis-factory mating – one bull acted as a psychological castrate whose presence was important to spur the dominant bull into mating. Australian research[32] found that on range, one bull on its own became active after midday while two bulls began their activity in mid-morning. The nature of this social facilitation has not been well studied. The presence of too many bulls, if evenly matched, may however lead to more fighting than mating.

Bulls and semen collection

For artificial insemination (AI), bulls have to be trained for semen collection. They may also have to be retrained after a lay-off period as this involves large changes in their environment. At collection centres bulls may be tethered to a wire, transported to and from collection points, required to mount teasers (cows, other bulls or mechanical devices) and so on.

Bulls and handler safety

The bull is recognised to be one of the most dangerous animals on the farm, and should never be trusted. It is clear however that the system of management under which they are kept has profound effects on their temperament, and this is often not taken into consideration. Few problems occur in young bulls and it is not until the bull reaches sexual maturity and becomes territorial that problems arise.

A German study of bulls at 29 AI centres[34] summarised the reasons why bulls attack men. These were:

- Lack of self-assuredness of the keeper who did not demonstrate dominance over the bull.
- Insufficient social contact between man and the bull.
- High-libido bulls were most aggressive.
- 77% of bulls that attacked men will do so again.
- Three-year-old bulls and those 7–10 years of age were most aggressive.

The bull's behaviour was improved by changing the environment, avoiding disturbance during sexual activity, using experienced handlers whose actions and clothes the bull recognised and not allowing previously attacked people near the bull. The bull's unacceptable behaviour was reprimanded immediately it occurred.

It is generally accepted that dairy bulls have a more aggressive temperament than beef bulls and that there may be breed differences. However, feeding, management and environment confuse the situation.

With bulls in AI centres, having to tease and frustrate them to improve semen quality prior to collection increases the risks of injury to handlers. Old bulls are usually most dangerous in yards or corrals, especially when they fight with other bulls and men get injured in the process. Bulls can be very agile and react quickly despite their size.

Bulls which run with cows or even groups of other bulls pose fewer problems than bulls housed on their own where they cannot see what is going on and are not allowed adequate exercise.

Bull handlers should clearly recognise the various body postures of threat and aggression so that appropriate action can be taken.

If cornered by a bull, it is best not to move too fast but to back away out of the 6 m range, watching the bull all the time until a fence or safe retreat is found. Turning and running invite being chased.

It is advisable for handlers to carry two sticks when handling bulls to extend the arms and provide a large visual image. The handler should recognise the point-of-balance of the beast that lies behind the shoulder. By using this point-of-balance the beast can be moved without going within its 5–6 m fight-or-flight distance. An escape path or man-gate should be built into handling facilities. A good dog that can heel and nose cattle should be near.

Bulls should not be held solely by the nose ring as this can tear out. A proper head halter should be used with a rope passing through the ring and two handlers involved if the bull's temperament is suspect or other people are at risk (e.g. at public displays).

Bulls and territory

A very important aspect of bull behaviour is how a bull becomes more territorial with age. Studies[22] showed that groups of 40 young bulls (under 2½ years old) set up regular grazing cycles with 60% clearly oriented in one direction during grazing. Social encounters were generally amicable and Jerseys were more active than Friesians. By 3½–4½ years old, vigorous agonistic and mounting behaviour started and a dominance order was clearly established. Certain areas were used for display and threat behaviour where holes were dug in the field. By 5½–6½ years old, the bulls were independent, very territorial and dangerous to shift from their regular paddock positions. Agonistic tussles and ritual threats occurred at the sparring holes which were dug in elevated positions if available to enhance the display. If space was adequate, bulls stayed in their areas without danger to each other. Problems only arose when they were brought together for movement or routine handling in yards.

Teaser bulls

Bulls can be made infertile by a number of operations:

- Vasectomy (tying off the *vas deferens*). Here animals can be done at any age, are completely sterile if done correctly and allowed to rest after the operation and before use. One testicle is some-times removed as a visual indication of their status and prevents their sale as entires. Because they actually mate the cow they can spread disease.

- Epididymectomy (tying off the epididymis). This is not favoured as the tubes can rejoin after a few months and fertility can be restored.
- Surgical deflection of the penis. Easily performed on a young bull which will be ready for service in 3 weeks after the operation, retains libido and works well through the frustration of no sexual contact. Does not spread disease. However some bulls do learn to serve cows.
- Penectomy (relocation of penis and urethra to the escutcheon). The bull cannot serve the cow. Such bulls will last for 2–3 seasons. The operation is costly and urethral blockage is common.
- Penile adhesions to prevent protrusion. These animals work well because of frustration of not being able to serve. It is expensive and the animal must be isolated during post-operative surgery.
- Induced cryptorchidism. Here the testicles are squeezed back into the body cavity and the scrotum removed by a rubber ring. Not recommended as bulls can be fertile and can spread disease.

In the experience of the authors, vasectomy is a satisfactory technique and avoids the need for excess mutilation. However, bulls should be done as yearlings, thoroughly checked for infertility before use, and have one testicle removed as a visual sign to prospective buyers. These teasers should only be used for one season, and where a mating harness is used, the teasers should be smaller than the cows. Dairy breeds seem to work better than beef breeds.

The castrate (steer or bullock)

Castrates are much easier to handle and are generally free from behaviour problems. In partial castrates (cryptorchids), although all male hormone levels are reduced, there is no evidence for decreased aggression or a reduction in other severe male problems.

The female

The female reaches puberty around 8–12 months of age but some dairy animals can show oestrus under 6 months old. The cow has an oestrous cycle which varies from about 18–23 days. Oestrus varies from 6–23 hours depending on the season, feeding level and presence and sex of other cattle. The effect of increasing herd size on oestrus is very important. Results[28] are described in fig. 5.

Signs of oestrus

A cow coming into heat may show some of the following behaviour. She may be:

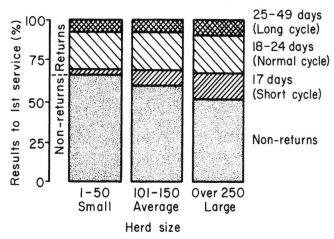

Figure 5 Changing patterns of short and long oestrus according to herd size[28]

- Restless and start walking the fences so eating and resting behaviour are disrupted.
- Found on her own or with cows not normally her associates.
- Much more aggressive towards other cows with many head-to-head encounters.

Approaching standing oestrus she may:

- Stand close to other cows, with chin resting on their backs, trying to make mounting attempts.
- Sniff around their tails, urinate frequently and butt away other cows.
- Show swelling of the vulva and have a full udder by withholding her milk.
- Encourage the formation of small groups with other cows.
- Walk around the field making lying cows stand up.

During standing oestrus the cow will:

- Stand to be mounted.
- Be covered in mud on flanks, have saliva on her back and clear mucus running from the vulva onto her tail.
- Make a lot of noise and walk around a great deal followed by other females and males if present.

Videotapes were used[19] to document some of the behaviour changes on the day of oestrus of cattle in barns (see fig. 6).

There has been debate about the diurnal or daily rhythm of heat expression in the cow. Workers[20] considered that the reason that visual detection seldom exceeded 60% was because most cows came into oestrus at night. However, re-drawing the figure and super-imposing the daylight management activities (e.g. feeding, milking) shows that these reduce opportunities for mounting to occur. Studies[24] in beef cows indicated that most activity took place during daylight though mounting was not precluded at night. This daily activity is essential if bulls are to be visually alerted to cows in heat.

Females – sexually active groups

In small herds, it is much easier for bulls to locate cows coming into oestrus perhaps days before 'standing oestrus' occurs. However, in large herds of dairy or beef cows with no bulls present, the cows themselves are usually seen in small groups revolving around the cows on heat which stand to be ridden by the others. These groups are defined as sexually active groups or SAGs[24].

In their study of 100 beef cows, these workers found two or three independent groups of cows with up to 6 animals in each. Above 6 animals the SAG tended to split up. During the feeding of hay or

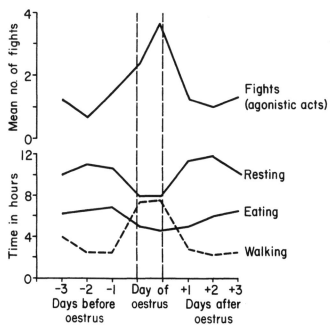

Figure 6 Changes in cow activity levels occurring on the day of oestrus[19]

silage, all visible mounting stopped but the SAGs stayed together. In the group some of the cows were riding but would not stand to be ridden as they were either at the very early or late stages of oestrus. Most activity in the SAG was between 2 cows both in oestrus. If there were odd numbers of cows in oestrus, then other non-oestrous cows would be associated with the group. Pregnant cows did not get involved in SAGs.

When bulls joined the herd and were drawn to the SAGs, all female mounting stopped. However the oestrous females did not readily stand for the bull until further courtship took place. Some courtship even seemed necessary before an oestrous cow would allow a female to mount her. The study showed a general pattern of cow–cow mounting to precede bull interest in the standing cow for 2 hours or more. Only once was an oestrous cow found by the bull first. Hence these SAGs have great practical importance in larger herds in alerting the bulls and the stockmen. It was also noted that some cows were less attractive within SAGs than others and gained fewer mounts.

Oestrus in stalled cattle

Cattle kept tethered in stalls or even in limited space in loose housing, especially during the darker winter months, are restricted in the outward expression of oestrus. This makes stockman vigilance very important in looking for restlessness, bellowing and reduced milk production. The greatest chance of making accurate observation is during exercise periods when the animals are free in a larger space.

Oestrus detection methods

Much more work needs to be done in oestrous behaviour before the best detection methods can be specified for each farming system. Currently, the best advice[46] seems to be not to mix systems and only to use one at a time until it is fully evaluated for each set of conditions. The systems currently in use include:

- Visual observations.
- Use of teaser bulls fitted with a marking harness.
- Progesterone levels in milk.
- Training dogs to sniff out cows in oestrus.
- Pedometers to measure distance walked.
- Painting the tail and rump and noting when this is scuffed off after riding (tail paint).
- Androgenised females, i.e. cows treated with androgens (male

hormones). It is not certain whether their behaviour is cow-like or bull-like and this will lead to differences in the reactions of the SAGs.

- Nymphomaniac (buller) cows. These can be used but may cause confusion and become physically exhausted. There is no documented information on their value.

There are many practical problems with marking harnesses. It is important to make sure that straps do not chafe the bull. If they are too loose the bull will cast the harness or pull it off on a fence while scratching. Regular checking to see that it is full of ink is also essential, and that the ink viscosity is correct for the climate.

Most errors arise in the interpretation of the marks on the cows, deciding which indicate a proper mount and which a yard mark or rape. Harnesses often work better on bulls which are smaller than the cows.

Oestrus problems

If a cow has not been observed in oestrus, it is often assumed that she has had a silent heat. The problem here is to determine if it has been a silent heat (e.g. ovulation without showing oestrus), whether it is a problem of poor human observation or the cow has not cycled at all.

It is important to realise that variations can be related to behaviour. For example, in field trials a cow with a long 15—hour oestrus was only mounted twice by other cows and was clearly less attractive. This happens with high-dominance cows. They may move about and pester others who are reluctant to mount. True silent heat problems have a low incidence in most herds.

In large herds oestrus detection is often made more difficult because the animals on heat cannot be identified in large groups milling around mounting each other. This is made worse when the cows are crowded in laneways, collecting yards or in indoor loose housing. Hired labour may not have the same motivation or experience as the owner and this can lead to poor results.

Research[29] into the occurrence of short oestrus cycles within 8—10 days of insemination showed that these increased as herd size increased and was most common among second- and third-calving cows. These short cycles occurred in herds with both entire and vasectomised bulls running with the cows so the problem was not caused by behavioural sub-oestrus.

With synchronised oestrus techniques, especially in large herds, accurate identification of cows on heat can be confused by 'riding chains' of up to 5 animals riding tail-on-tail. Activity in these herds reaches very high levels with resultant confusion. Some farmers use 'blanket insemination' (where they inseminate the whole herd) to make sure all cows on heat are mated.

Under extensive range conditions and with large herds where artificial insemination is used, operators rely on visual oestrus observation by stockmen constantly riding with the herd. They may also construct small corrals in the middle of the pasture where an entire or vasectomised bull is held. Two such pens may be arranged so that the herd has to pass them on their way to water. Females coming into oestrus or in oestrus are drawn to these bulls. Such cows are often identified by large numbers on their flanks.

Courtship and mating

For successful mating two things have to happen – the female escape responses must be suppressed and male aggression be appeased.

Figure 7 sets out the processes which occur before copulation. Both the male and the female give out signals that they are available for mating quite independently of whether there are other potential mates nearby or not.

Called 'broadcast stimuli' these are important and are seen in an isolated house cow which is more restless and bellows more

BROADCAST	IDENTIFICATION	SYNCHRONISING

Independent of presence or absence of others	Close quarter to potential mate	Type 1: Physiological preparation Type 2: Courtship, female and male in concert Type 3: Copulation, egg and sperm

Figure 7 The stimuli provided by male and female in the period leading up to mating (after Schein, M.W. and Hale, E.B. (1965). Chap. 18 in *Sex and Behavior*, ed. F.A. Beach, New York: John Wiley)

when in season than when not. Vocal calls are one of the main broad stimuli given off by oestrous cows but may be ignored or misinterpreted by stockmen.

Odours aid the sexual identification. Cow odours provide the clue that she is a female and that she is approaching or is in oestrus. Sniffing of the perianal region of the cow by the bull is common and this alerts his attention to the female in heat.

When a cow in oestrus is sniffed, she may also urinate. The bull licks and tastes the urine which stimulates the lip curl of flehmen response with head lifted and neck stretched out. As well as allowing the odour to filter into cavities above the roof of the mouth, the body posture signals to the female that the male is amicable and not aggressive (head-down posture).

The flehmen is given not only during the courtship responses of bulls to cows but is seen also in bull-to-bull and less often in cow-to-cow responses. Bulls often use the flehmen response when meeting other bulls on the way to water.

Once the male and female are together 'synchronising stimuli' enable them to be correctly co-ordinated before actual coitus (mating). In this period of readiness courtship occurs in the form of approach and chin-resting by the bull on the back and rump of the cow. Intention mounts without actually doing so and half mounts are common.

Finally, there is mounting and copulation. The bull clasps the cow with his forelegs and on ejaculation he may jump clear of the ground. After dismounting, there is a refractory or recovery period until a return of interest by the bull and he continues with his mating response.

Natural patterns of mating

An experienced bull running with a small group of cows where few will come into heat at the same time will single out the cow up to 48 hours before full oestrus onset. He will stay with her, graze nearby, lie next to her and keep a 'guard' over her till she is ready for service. As she approaches full oestrus the bull sniffs at her vulva and attempts to mount. There may be partial erection, dribbling of accessory fluids and finally when mounting is allowed it is performed quickly and the ejaculatory thrust is given with much vigour.

Other displays of masculine threat may be given to nearby bulls while the dominant male 'guards' the cow. This includes pawing the ground, throwing dirt over the back and withers and horning the soil while vocalising and threatening nearby animals.

Production

Grazing behaviour

A cow's day is divided into clearly defined periods of grazing, rumination (chewing the cud), idling and rest. If it is non-lactating, such as in grazing bulls, there are three main periods of the day when the rumen is filled with freshly harvested grass. These are daybreak to mid-morning, mid-afternoon to half an hour after sunset and a shorter period about midnight.

The picture is more complicated for the milking cow. Depending on the size of the herd up to 3½ hours/day may be spent around the milking facility, and milking may be set at times which otherwise would be used for peak grazing (see fig. 8).

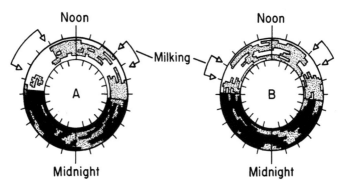

Figure 8 Typical weekly grazing patterns for a cow recorded by a grazing clock. A, second week of lactation; B, peak lactation when 4 weeks in milk. Grazing time = hatched; day = white background; night = black background

Dairy cows may be forced to increase their night-time grazing period from about 1 to up to 4 hours to compensate for decreased day-time grazing. This is especially the case when nutritional demands are high such as after calving.

The amount of time a cow spends grazing can be recorded by a grazing clock strapped to a halter on the animal's head. The swing of the pendulum in the clock is recorded as the cow's head moves through an arc of 60–80° when eating.

Grazing studies have shown that cows normally spend 8–10 hours a day to satisfy their appetites. However, this varies due to many factors such as weather, pasture quantity (length) and quality, the nutritional needs of the cow, oestrus, pregnancy and health.

The newly calved cow is in negative nutritional balance, i.e. her energy output is greater than her intake and she milks off body reserves. Also as a cow's uterus expands during pregnancy, rumen capacity is restricted. Reviews of this subject have been given in references 1 and 11.

A main effect of all these variables is seen in the length of time the animal spends eating. In lush spring pasture conditions for example, the cow may only need to graze for about 7 hours to satisfy appetite, and at the same time its nutritional needs. This contrasts greatly with dry tropical pastures where studies[44] reported from Queensland that Jerseys grazed for 9–12 hours/day. It was considered that 12 hours was the upper limit of grazing time and only under extreme pasture shortage had grazing times in excess of 13 hours been recorded. It was also noted in similar trials on these pastures that leader cows produced 38% more milk than follower cows, the latter also having poorer milk letdown.

Research[44] showed that grazing time provided a less precise measure of feeding behaviour than the rate of biting which varied between the grazing periods. Variation in biting behaviour was due to the ease of prehension of herbage, grazing bites and pasture quality. During the cows' 9–12 hours rumination and grazing, a total of 45–75 000 bites were involved of which a maximum of 36 000 bites occurred during grazing. Cows are capable of prehending less than 0.3 g of pasture organic matter/bite.

In tropical conditions with high daily temperatures, the main grazing periods are reversed and occur at night.

The immediate post-calving period causes greatest stress to the grazing dairy cow. First there is the stress of having its calf removed and then the move into the milking herd with new routines. Added to this are the high nutritional demands of lactation causing extended grazing periods. Supplementary feeding (if economic) can reduce night-grazing times.

Behaviour studies, such as those of reference 14, showed how cows act in unison in their grazing patterns. Hancock showed that two herds maintained their own unique grazing cycle, even when only separated by a single electric wire. He postulated that each herd operated under its own set of herd laws.

Social facilitation also operates and concerns some farmers. For example after a period of rumination and rest, the herd may be stimulated to re-start grazing by copying the first cow to resume. The production level of the herd may thus be controlled by whether the first cow to re-graze is a low yielder or a high yielder. A practical solution to avoid this is to split the herd on the basis of production.

Feeding

In many dairying systems in different countries, concentrates are fed to cows during milking. There are many reasons for this based on practical convenience and economics. Thus feeding becomes part of the conditioning stimuli to milk letdown. Problems may arise when this feeding is stopped as letdown may be affected. To avoid this, feeding should be irregular so that the cow does not expect feed each time it is milked. This advice is also given regarding feeding supplements to cattle in drought conditions on range. The aim is to ensure that the extra feed is viewed by the cow as a supplement and not a substitute.

The food preferences of cows have recently been studied by using animals trained to press a nose plate to obtain food of their choice in a two-choice situation[26]. A range of by-products was tested, with cows offered only limited amounts so that satiation was not a factor. Figure 9 shows the ranking of 20 feeds.

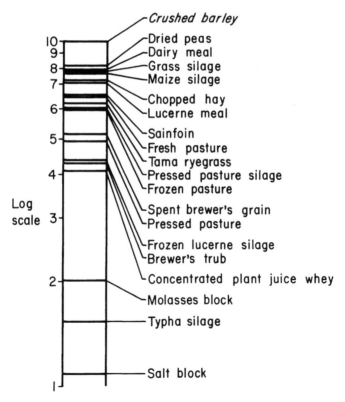

Figure 9 Ranked preference order (10 = highest) of twenty cattle feeds[26]

The feed falls into three basic groups:

- Green fodder or more traditional feeds.
- Average – preferred products which have been processed.
- Non-preferred products.

This scale forms a base against which new feeds with potential use can be tested and problems of acceptance assessed. Although there are a number of flavours which are supposed to increase the acceptance of non-preferred foods by cows, to date there is little published evidence to prove that these flavours achieve their claims. There is a need for effective masking flavours to assist cattle to eat non-preferred by-products.

The food preference work has highlighted the ability of cows to learn to operate mechanical devices which encourage them to work for their feed. This could be exploited to save labour, reduce problems that occur in excessive idling time and allow stock to administer their own supplements and drugs. The drum-lick roller is an example of this.

Control of feed intake

The main practical behavioural problems arise where the stockman wants to vary the animals' nutrient intake. There are many ways of doing this and some are listed below:

- In pasture grazing, use an electric fence to alter grazing pressure through allowing more or less land area and hence dry matter offered.
- Feed by bulk feeds such as hay, root crops or straw to satisfy appetite, before allowing concentrates.
- Feed a complete diet where all the components are ground up and pelleted. This allows more control of the animals' individual needs as opposed to offering a variety of feeds *ad libitum*.
- Increase the intake by improving the quality of pasture or crop by-products. Cows will eat more hay if it is cut before flowering and is green and leafy. They will eat more silage if cut before seeding and properly fermented.
- Feeding the diet in smaller amounts more frequently, and making sure stale feed is removed and troughs and bunkers are clean.
- Feed additives which alter the flavour of feeds can be used. Cows vary in their individual preferences.

- Alter the physical form of the feed mechanically by chopping up roughages or treating straw with chemicals.
- Avoid damaging pelleted feeds by mechanical devices which can produce unpalatable fine dust (fines).
- Watch for augers and conveyors that may not distribute the feed evenly along the troughs in cattle sheds or feedlots. Animals at the very end may be underfed.
- Remember that individual electronically controlled head bails may not be foolproof for some animals which soon learn to steal from other cows.
- Check the trough space or space at the face of the feed (e.g. silage). This should be calculated to allow a pre-determined average intake/animal.
- The amount of time the cows are allowed access to feed can be manipulated. Social rank is vitally important here, as low-rank cows may never be adequately fed.

Rumination, idling and drinking

After eating, the ruminant as an 'eat-and-run' animal retires to a place of safety to ruminate. Here boluses of partly digested material are regurgitated, chewed to finer particles and then re-swallowed.

Observations show that about 65–80% of rumination occurs while the animal is lying down, though in wet weather the cow will ruminate more while standing. The animal's adult rumination patterns are set by 6 months of age and normally 15–20 periods of rumination are scattered throughout each 24 hour period. Up to 1 minute is spent chewing each bolus. Feeding concentrate feeds (grain) to cattle in contrast to fibrous feeds reduces the time spent ruminating.

The amount of time spent idling (i.e. standing still and not eating or ruminating) varies greatly between animals. Cattle in feedlot where feed intake is rapid spend more time idling than grazing stock. It is often during these idling periods that management problems arise such as riding in bulls and bark chewing of trees. It has been suggested that finding ways of extending grazing time could reduce these problems.

The grazing ruminant is designed to walk considerable distances and this is very obvious from studies where vibracorders have been used to record grazing and ruminating times in the field[36]. In small fields the cows' activity differs greatly from that in large fields where both walking and loafing initially increased due to the novelty of the new environment.

In extremely cold conditions on winter range in Utah it was

calculated that on cold days cows reduced their activities and saved 14% of the energy normally expended on warm days. They saved up their activities such as food gathering for the warmer days. They demonstrated a behaviour pattern designed for efficient energy use.

During drinking cattle dip their muzzles into the water and suck the fluid. Their water requirements vary widely depending on climate, quality and quantity of feed intake, and production status. High producing dairy cows require up to 160 litres/day, and cattle on high protein diets can drink 26% more water than those on low protein diets. Observations show that water is drunk between grazing periods at pasture and often in the collecting yard before milking. Cattle are often given dietary supplements and bloat preventatives via the water supply but intake can be variable.

Sleep

It is generally accepted that adult cattle sleep very little. A typical sleeping pose is when a cow lies with her four legs tucked underneath her, chin resting on the ground and head turned round to face her rear.

Some observations[35] on 2 cows showed that sleep pattern was not altered during pregnancy until 24 hours before calving when total sleep time was halved. Sleeping pattern was also altered by any general upset such as calf removal and housing cattle after being at pasture. Here it could take 3–6 days to establish the normal sleep sequence. It also took 3–6 weeks to establish stable base values for stall conditions. It was also observed that in stalls sleep was restricted to night time and when cows were restrained by stanchions as opposed to tethers, it took 4–10 days for the animal to find an adequate position to re-establish her normal sleep pattern. If animals are disturbed at night, they will sleep more during the day.

The important practical point is that unless the animal is well settled and comfortable it cannot sleep. Changing cows into strange stalls, changing their routines, and uncomfortable stalls will all have an adverse affect on production and cow well-being.

Housing

The degree to which cattle are housed varies greatly throughout the world. In certain parts of Europe cows may be tied in stalls all their lives, while in Australasia they are always at pasture. Careful tree planting to provide shelter for animals should be exploited more on farms. This could be used for calving and allow stock to stay outdoors and help reduce the capital cost of housing. In American feedlot systems they may be loose-housed in fully, partly or non-covered lots.

Behavioural problems can arise in systems where cattle are tied indoors. They are due to the pressure to reduce labour involved in feeding, watering, bedding and cleaning out. As well as this, stockmen have to be vigilant to detect oestrus and disease. In these situations, small numbers of cows/man usually ensure higher husbandry standards.

In total confinement indoors (loose-housed or stalled) dairy cows still show similar behaviour in both duration and distribution to cattle at pasture. The periods of eating, drinking and social activity are most intense during and shortly after milking and feeding times.

Studies in Sweden[16] of tied dairy cows highlighted the following points:

- The standing area must be designed to fit the cow's behaviour needs.
- The stall area must enable her to stand and lie in a natural way, allowing free vertical movement of head and neck to get up and lie down.
- Dimensions for short standings should be 1200 x 1700 mm and 1200 x 2200 mm for long standings.
- Partitions are of great importance for keeping the cow in her place and preventing trespass. Partitions should be between every second cow. This will prevent trampling injuries.
- Feed racks should not hinder the cow's head movements.
- The cow should be able to reach all parts of the feed trough and water bowl with ease. Troughs should not hinder the cow lying down and their design should keep the food within it when the cow is eating.
- Certain adjustable feed racks often overtire the cows, as expressed by a higher lying frequency.
- Chore work should be designed to suit the cow's normal behaviour and irrelevant visits and disturbance of the animals should be avoided.
- Cows should not associate the entry of people with feeding, but stay resting.

The main concern is to make sure the cows are comfortable so that problems do not arise. Cow comfort needs as currently known are shown in fig. 10. In this regard, attention should be paid to

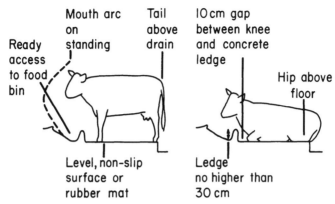

Figure 10 Stall design features to meet the need of indoor tied animals (after Wander, J.F. (1976). *Zuchtungskunde* 48(6), 447–59)

providing an adequate environment for both cows and workers. The provision of shelter is important, especially in areas where cows have long periods standing waiting in collecting areas. Recommendations are for a temperature of at least 7°C for dairy cows with less than 80% humidity[51]. Draughty conditions often cause chapped teats and increased disease incidence.

Most problems arise with cows in stalls that are the incorrect size. This increases the work in cleaning, causes feet and leg problems and results in trampling injuries to other cows, especially the udder and teats. A solution is to have stalls where the standing platform can be lengthened as the animals grow. This is done by adding sections of grating at the back of the stand extending into the dung channel. Certain older high-producing cows are often kept in a separate loose box to ensure maximum comfort.

Many farmers like to let their cows out once a day for exercise, heat detection and to allow cleaning of the cowshed, providing feed and bedding. This also allows normal social interactions to take place among the members of the herd.

The removal of dung is the least popular part of the farmers' chores and various devices are used to encourage cows to defecate and urinate in the channel behind her and not on the standing so that her back feet then stand in it. An electrified trainer hangs above the cow's back which discourages her from standing forward to evacuate. As she arches her back, she contacts the trainer which forces her to stand further back and eliminate in the channel. Research[4] has shown that cows can be trained to evacuate at particular times. Cows usually defecate after they have been lying and this behaviour can be exploited.

In complete loose housing, cattle lie on an area bedded with straw, shavings, sawdust, industrial wastes or in dry climates perhaps a mound of earth. With adequate feeder space, cows can behave as a normal herd.

In other systems cubicles are designed as stalls. There are also areas allowed for loafing when animals are not at the feeding face or resting. Behavioural problems arise where some cows refuse to lie in the bedded cubicle and prefer to lie in the dirty loafing area. The ratio of the number of cubicles to the number of cows is critical, as there is usually a high peak demand for cubicles in early morning. Recommendations vary depending on herd size and the system. For herds of up to 40 cows, suggestions are to have a cubicle for each cow, but in larger herds there could be 105 cows to 100 cubicles. Other suggestions are to have 15–20% fewer cubicles than cows. Clearly the stockman should decide on a final ratio after observation of the herd's behaviour. Provision should be made in the original plans for later modification.

The cow's lying preferences have been studied. In free-stall barns a slight slope of floors tends to encourage cows to lie with their backs in the same direction. They would then fit into the stalls better and there would be less treading injuries to teats.

The behaviour of housed dairy cows has been shown[39] to depend on the level of feeding provided. Cows receiving low planes of nutrition spent most time in non-productive activities such as idling, licking, eating metal rubbish (hardware) and walking, while cows on high planes spent more time eating, standing, ruminating and defecating. Confined cows walked only 610 m/day which is about one tenth of the distance travelled by range cows.

The behaviour of steers in feedlot has been reported (see reference 17) from both summer and winter observations. With self feeders, total daily eating time was 2 hours 20 minutes. In summer greatest activity occurred in late afternoon (3–6 p.m.), early evening (6–9 p.m.) and a smaller peak near sunrise (6–9 a.m.). In winter greatest activity occurred in the late afternoon (3–6 p.m.). Greater periods of time were spent drinking and in the shade in summer than in winter. Cattle spent 12 hours/day lying in both summer and winter which contrasts greatly to the behaviour of the active range beef animal[47]. See reference 37 for further information.

The cow is a cold-climate animal, and is more stressed by heat than by cold. The question should be asked why so much capital is invested in elaborate cattle housing. It often appears that the housing is more for the comfort of the worker than the animal.

If housing is simplified, many welfare problems disappear. For cattle, more feed and the stock left outside may be a cheaper alternative than providing a building. This should be considered.

The size and shape of pens has been based more on economy of construction than on cattle needs. Studies[45] showed that pen shape affects the way steers stand with respect to the pen and its features as well as to pen mates. Steers in groups of two or five tended to keep their heads more at the perimeter of a pen than at a centre; so the centre was used inefficiently. On this basis triangular and square pens would provide more effective space for the stock than a circular pen.

Leadership and social dominance

Dairy cows move in a consistent way or leadership order. Mid-dominant cows tend to be in front of the group although seldom are single individuals found as consistent leaders. Instead, there is a pool of animals which tend to be in or near the lead.

There is more consistency found in the cows at the rear of the moving herd. Hence 'rearship' is a more distinctive feature than 'leadership'. Animals at the rear are usually the younger subordinate heifers.

The distances that cows walk to be milked and the nature of the walkway need special consideration by stockmen to avoid injuries to cow's feet and udders. Small pebbles or sharp stones on walkways cause feet injuries. In some countries cows in large herds are now moved by men on motorcycles and are often made to trot instead of walk. This adds further stress.

In most herds there is a definite order in which cows enter the milking parlour. Again the mid-age, mid-dominant cows tend to be milked first in herds of up to 100 cows followed by the older cows[40].

Social dominance orders (called 'bunt orders' in polled cows and 'hook order' in horned cows) become more complex as herd sizes increase. Once established they remain fairly stable. In studies with identical twins split in two herds, twin mates tended to have a similar dominance ranking within their own herd.

Social dominance must be considered in designing facilities for the herd. An example is the construction and placement of water troughs. Cows prefer to be able to see while drinking to avoid being butted, and more can drink at once from long narrow troughs than from squat round troughs. Round troughs are more efficient when placed against fences rather than in the centre of paddocks but still provide water for several cows when split by the fence between two fields. In dry conditions several smaller troughs in the same field would provide better watering for large herds and at little extra cost. At pasture it is important to have adequate water (space and flow) to allow the herd to drink as a group activity after they finish grazing as a group in drought periods, otherwise production can be affected. In systems where cows are kept in loose housing or feedlots, the trough becomes an important area for interactions between animals.

Some practical points related to social dominance effects are:

- Cull the low-producing boss cows and weak submissive cows.
- Dehorn all cows as calves.
- Do not shift cows to different groups. Try to keep group size stable and no larger than 100 cows.
- The slope to the feeding floor in front of the trough should be about 1–2%. Cows are frightened to walk on steeper slopes which become greasy and smooth. Remove slippery film with muriatic acid and roughen the surface.
- Adequate feed-bunk space is important. For heavy corn silage diets, for example, 60 cm/animal is ideal.
- After milking let the cows return to fresh feed. This delays the cow lying down till the udder is dry.
- In large groups, watch out for cows that may be under 'social pressure'. They may need special care.

Social harmony in a group of cows is extremely important to maintain good production. A few suggestions to avoid social instability are these:

- Reduce calving spread so that early and late calvers do not have to be separated and then re-united.
- Keep calved and uncalved cows in adjacent areas so they can maintain social contact across a fence. This would be effective for short periods.
- Alternatively, the two groups could be brought together for a short period daily or weekly to reinforce stable dominance relationships.
- Maintain cows as one group throughout calving so regardless of their lactational status they all go through the parlour together.
- Calving heifers before the older cows will allow them to establish their dominance relationships and to have become familiar with

the milking routine before the older dominant cows joined the milking group.

Large herd problems

Experience in many countries leads to the recommendation that whatever the size of the total farm herd, cows should be run in subgroups of about 100 animals. There are many examples of production increases when herds were split such as a 10% increase when 120 cows were split into two batches of 60 at Reading University. Other farmers consider 70–80 cows an optimal group.

The problems are mainly concerned with:

- Excess crowding.
- Cows getting down and being trampled without being noticed.
- Very poor oestrus detection.
- Long periods spent standing on concrete.
- Problems with boss cows at water points and feeding facilities.
- Great problems with the bullying of heifers and if separated, even greater problems when they were reintroduced to the herd.
- Poor milking management.
- Poor vigilance for disease incidence.
- Stockmen working under greater pressure in larger teams hence less job satisfaction and reduced attention to detail.
- Greater disease incidence in the cows caused by the stress of the system.
- Large herds need extra labour which may lack incentive, leading to poor husbandry.

Milking

Dairy cows must be trained to the milking routine. In developed dairying countries with genetic improvement programmes, dairy temperament has been selected for and large gains have been made. Key aspects of dairy temperament importance are standing quietly while being milked, having the udder handled without kicking, avoiding dunging or urinating while being milked, having good 'let-down' with minimal stimulation, milking out quickly and so on.

The most important part of the milking routine is to stimulate the milk 'let-down' response of the cow. Cows can be conditioned to a wide range of stimuli associated with the milking routine and even to such devices as coloured discs shown to the animal at milking[50].

In countries where breeds based on *Bos indicus* are used, there are still problems associated with milk let-down. These breeds need the presence of the calf for good milking stimulation. The technique is to let the calf suckle the cow for 2 minutes before milking or allow the cow to see, lick and smell the calf while being milked.

In developed dairying countries there has been great progress through selection for dairy temperament. In New Zealand in the 1930s cows had to be stimulated by washing and massaging and hand stripping was used after machine milking. In the 1980s cows will let-down their milk at the first touch of the machine or when walking into their milking position.

Milker temperament

The temperament of the milker has been recognised as important from very early times. Most recent scientific studies[42] have documented the important relationships between cowmen and cows.

In a study of 28 herdsmen with identical policies, calving dates etc., the cowmen's attitudes, motivations, personalities, actions and behaviour were analysed and related to average annual milk yield/cow.

The main findings were:

- Cowmen with higher job satisfaction achieved greater yields/cow.
- Cowmen sought different achievements from their work, were motivated differently and had different aspirations.
- Herd performance and cowmen's age and status were not related.
- Confident, introvert cowmen achieved greatest yields with identical inputs and conditions. Figure 11 shows a general hypothesis from these data.

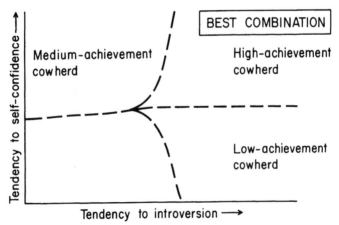

Figure 11 Personality traits of cowmen and their effect on cow production[42]

- Non-milking activies (general farm chores) are rated as less satisfying than milking activities. Cow-contact hours should be high.
- A period of highest stress occurred when one third of the cows remained to be milked. Cowman stress declined as the end of milking approached.
- Two or four women are said to be better as a milking team than three women.

This information[42] has been used to teach cowmen how to achieve a better relationship with their cows and hence increase yield. It is based on the cowman developing a greater awareness of the cow's behaviour and being generally more observant. The following is a list of examples of what cowmen should look out for:

Location of cows	Possible changes in behaviour
In collecting yard	Abnormally active
Entering parlour	Cows that are usually willing to enter now refuse Order of entry differs
Being milked	Loss of appetite Low yield Normally quiet becoming restless Holding milk compared to normal
Going to pasture	Cows in different position in moving group Cow not moving with rest of herd
At pasture	Abnormally active Not grazing with rest of herd Standing with head down
Winter housing	Cows that normally lie in cubicles lying or standing in passages Cows not at silage face when normally there

Many milkers have the radio playing during the milking routine and it is suggested that this directly assists milk let-down. It is more likely that it relaxes the milker whose behaviour in turn affects the cows and is seen in their behaviour and milk production. Similarly talking to cows in a quiet pleasant voice and patting them will achieve a good atmosphere in the milking parlour and reduce dunging. Sudden noise or shock to the cows during milking usually results in a stop of milk flow and increase in defecation. A loud radio as a background noise certainly reduces the impact of other sudden noises.

The milking machine itself is the most important part of the entire milking facility and it must be serviced and checked regularly. It is the main source of mastitis spread and money spent on its maintenance will reduce that spent on antibiotics. High-pitched noises from the machine not heard by man can be heard by cows.

Training heifers

The newly calved heifer that has to be milked for the first time needs special care and consideration as her early experience of the routine will be remembered for a long time. Apart from putting her through the parlour on a regular basis before she calves and washing and massaging her udder, it is important at her first milking to:

- Make sure the cups do not squeak.
- Not let the cups fall off.
- Keep her head up.
- Make sure heifers on a herringbone parlour are tightly packed.
- Avoid overmilking.
- Never 'lose your cool'.

Some heifers still prove difficult and various methods are used to get them to stand. Some examples are:

- A tight girth rope in front of the udder which makes her unable to kick.
- Leg ropes to tie the leg nearest the milker to a part of the parlour.
- Rubbing the pit of the heifer's back to soothe her.
- Lifting her tail to a near vertical position or pulling it down firmly. This can injure the tail. Care is needed.
- A metal clamp which goes across her back in front of the hips.
- Nose grips and pulling the head round to face the rear.
- Blindfolding by putting a bag on her head.

If such methods have to be used, then the entire handling routine needs careful examination starting with the calf.

Parlour design and cow flow

There are many variations in the design of milking parlour and new developments are constantly occurring. In making a choice the farmer should pay special attention to the needs of the cow as well as the cowman to maintain a happy relationship between them. There are many behavioural aspects that need further study such as the importance to the cow of standing in a standard-sized and standard-shaped environment on a rotary platform compared to being pressed against other cows in a herringbone. Time and motion studies have highlighted that the time spent loading the cows into the milking facility is the greatest problem area in modern operations. The exploitation of cow behaviour is of great assistance in avoiding resort to mechanical aids and physical force.

Cow flow has many important implications in the milking routine. For example where higher-yielding cows come to be milked early, these are generally the more dominant, larger and older cows which also have a greater incidence of udder infections. The infections can be spread via the machine to the non-infected heifers that are milked later.

In most milking facilities, the cow flow towards the parlour is interrupted by the herd waiting in the holding yard. From there, cows are brought forward as small groups or individuals to the milking positions. This interrupts the initial spontaneous flow of the herd from the field and fails to recognise the importance to the milking speed of a continuous flow process.

To overcome the block in flow, devices such as backing gates or crowd gates, large revolving comb structures, water jets, electric wires and physical beating are used to get the cows into the milking position. The danger with these is that they encourage a negative response from the cow and prevent milking being associated with pleasant routines.

Opinions have varied over whether cows enter a milking parlour at random or in a set order. Studies[12] on 3 British herds (56, 57 and 104 cows) showed that 80–85% of the cows entered the parlour at random with 15–20% entering in the same consistent position. As lactation proceeded, the cows became more consistent in their entry order with a slight tendency for higher yielding cows and those in early lactation to enter earlier than older cows.

Problem solving

Investigations of problems that appear to be in the parlour or with the milking routine should be examined in the reverse order to which they occur. The investigator should go along the cows' path in reverse order checking various points. Areas worth investigating are:

- Width of the walkway to and from the parlour.
- Walkway surface – rough, slippery, muddy.
- The exit from the milking position – blind corners, floor surfaces.
- Electric shocks in the parlour.
- Patterns of filling the parlour – cows refusing to move to front bails in herringbone.
- Water jet systems for cleaning the cows and stimulating the udder.
- Crowd gate – shape, surface.
- Crowd gate – operation, fast or slow, electrified.
- Blind corners to block flow.

It is a good exercise for the stockman to examine the unit from the cow's eye level and assess the milking routine in terms of positive and negative responses for themselves and their cows. This can be done as a balance sheet. Examples of negative points are where cows are keen to get out of the parlour before the exit gap is large enough, the number of times the milker has to chase a cow into the milking position, the amount of dunging in the milking area and the kicking off of machines. Sometimes small changes can bring about large improvements.

Points of good routines

The following is a summary of some points worth integrating into any management routine with dairy cattle:

- Handle calves from an early age.
- Move quietly among stock and talk to them at every opportunity.
- Run in-calf heifers through the milking facility with the lactating cows to get them used to the routine.
- Be patient when training heifers to the milking routine.
- Keep milking routines and milking times consistent.
- Prepare the cows quietly for milking.
- Use milkers with a quiet temperament who like cows.
- Organise the system so that milkers do not have to milk for longer than 1½ hours.
- Move cows quietly to the cowshed and into the milking facility. Don't rush them.

- Avoid backing gates that are electrified.
- Get rid of sticks, electric goads and dogs around the milking shed, and stockmen who persist in their use.
- Have all the milking equipment checked at regular intervals, especially pulsation.
- Hold cows for the least time possible in the yards.
- Veterinary examination and treatment should not be done within the milking parlour.
- Check the milking bail for correct earthing to avoid low voltage leaks reaching the cow through the teat cups.

Electronics on the dairy farm
The use of electronics in the identification of cows, measurement of their feed, their production, liveweight and health status will increase rapidly in future. These developments should be encouraged provided they increase the well-being of the cow and allow the stockman more time in observance of his stock.

Dung and its problems
Cattle defecate at any time throughout the day and in any place. There is a greater chance of defecation after standing or lying for some time, if they are disturbed or frightened by unusual stimuli. Operation of crowd gates in the holding yards, especially if electrified usually stimulate excess dunging.

Although there are some peaks, there is a fairly consistent faecal output during day and night. In the 1950s research[15] estimated that about 5–6% of the dung from a dairy herd was dropped near the parlour. As herd sizes increased in the 1970s–80s and more concrete laid around parlours this percentage has risen steadily. Engineers currently allow for as much as 25% of a cow's dung output per day to be left near the milking facility.

Formerly, the use of different fields for day and night grazing led to fertility transfer between day and night paddocks. With milking times at 6 a.m. and 3 p.m. more dung and urine was left in the night paddock and less grazing was carried out. However with more dung now left around the parlour, this problem needs further investigation, especially with the rising costs of artificial fertilisers. The disposal of eliminated wastes presents a major physical, environmental and cost problem to feedlots and intensive units[37]. The energy cost of moving this material and the water required often outweigh its value as a fertiliser and reduce the profits of the enterprise.

Many studies have confirmed that cattle avoid grazing herbage contaminated by dung, and it is generally agreed that taste and smell are involved. In one trial where rejected herbage was eaten by heifers when fed to them indoors, it was concluded that smell was the main factor.

In a review of the subject[31] it was suggested that utilisation of dung-affected herbage could be increased by:

- Accelerating decomposition. This traditionally has been done by harrowing but it may cause damage and make more of the pasture unpalatable.
- Increasing stocking rate and hence grazing pressure depriving the animals of a choice of herbage. Here animals will graze nearer to the dung.

In some farming systems where slurry (diluted dung) is produced and disposed of by spreading on pastures, grazing problems can arise. Research[33] for example showed that cows on slurry-covered pasture spent longer time grazing, walked about more during grazing, took smaller bites of grass, engaged in competitive interactions more frequently and drank more often than cows on pastures where the slurry was injected into the soil.

A Japanese experiment to condition cows to graze disregarding their own dung was not successful. Conditioning was done by attaching sponges soaked with skatole (the decomposition chemical in dung) above their noses. This product of dung might be useful as a repellant. However, the use of skatole to deter animals from de-barking trees was not successful, while dung itself was[30].

Training cattle to lead

Cattle may be required to accept a halter and be led for show or sale, and for veterinary inspection, treatment and transport. Training cattle to lead, even as young calves can be time-consuming and physically demanding. At later ages, especially with older stock such as bulls, it can also bring risks of injury to both the animals and their handlers.

To achieve good results the principles of primary socialisation are needed along with kindness mixed with dominance, reward, punishment, repetition and re-inforcement. It is important to realise that animals differ widely and part of the stockman's skill is to recognise the beast's temperament and how to exploit it.

The methods used to teach cattle to lead are as follows:

- The animal is first tied up to a solid object for 30 minutes/day for a week and perhaps groomed or handled so that it loses its fear of man. It can be fed at this time also to help in taming.
- Pulling the animal using a halter and long rope. The long rope allows greater leverage to stop the animal if it takes off. As it learns to lead when pulled, a shorter rope is used. Two or three people can help to pull initially while one person stands near the beast to quieten it. A person can stand behind the animal also pushing it.
- Using the same method as above but using a tractor to pull the animal. Again the handler stays near the beast to quieten it and let it associate the man with the pulling. A person may have to stand behind, goading the animal forward with a stock or electric prodder. These should be avoided if possible.

  In pulling animals, a knot should be put in the halter rope so that the slip knot does not tighten and injure the beast's jaw. Regular checks should be made for rope burns.

  Some stockmen try to arrange for the animal to fall over on to its side in the early stages of pulling. This often gives the beast a fright and it is less keen on resistance. This is very easy to do on slippery (icy) ground although stockmen will have similar problems in keeping their feet.
- Two animals of similar size and weight can be tied together with collars and a 500 mm chain with swivels so that one is pulled by the other. The animals are then led by people after the initial heavy work has been done. They can be led for a short period each day then chained together until they lead easily. A close eye must be kept on them during the first day or so to untangle them if needed and an open paddock free of obstacles is needed.
- Animals may be tied together that are not of similar size nor indeed of the same species. A large animal will quickly teach a smaller one to come when pulled and a donkey has been used successfully by one breeder to teach show cattle to lead.

Cattle can also be taught to stand correctly for showing and photography. This is done by using a stick or show cane to touch the feet until it lifts them and puts them in the correct position. The head can be held up and its attention attracted by a noise or a visual distraction. Cattle respond to kindness and retain a memory of bad treatment for a long time.

Welfare

From the current information available on cattle, it appears that there are certain key areas which need priority when considering welfare within a production system. These areas, along with points to consider are:

Calving and the calf

- Recognise the time course of the suckling reflex and make calves work for their milk when reared artificially.
- Have plenty of teats available to allow for social feeding in grouped calves.
- Colostrum is critical. All calves must have it as soon after birth as possible.
- Calves need shelter in the first few weeks of life.
- Young calves should be handled and transported with care.
- Infrequent feeding is not necessarily stressful. It imitates the natural patterns of feeding.
- If possible calves should be removed from the dam at the moment of birth to minimise stress. This is rarely practical unless labour is plentiful.
- Grooming is important to calves. Provide it where it is practical.

Stalls and indoors

- Calves are a lying-out species that appreciate isolation in early life. Thus individual stalls provide a good environment for the first week of life. Thereafter calves require social contact.
- Draughts should be eliminated for indoor stock and bedding should be provided if possible. Cattle appreciate comfort.
- Stall designs should be flexible to cater for the growing animal.
- Animal comfort should be the top priority in stalls. The animals' needs should come before those of the stockmen.
- Shelter and shade are important to cattle. Systems should consider them more, especially when trying to reduce the high capital cost of housing.
- Systems (both indoors and outdoors) should be based on the beasts' needs.
- Only comfortable cattle sleep and produce well.

Bulls

- Treat all bulls with caution, especially those 3 years of age and animals 6 years and older.

- Lead bulls with a halter, without putting pressure on the nose ring.
- Recognise that bulls are territorial animals.
- Appreciate the 6 m threat distance for bulls.
- Keep bulls interested. Make sure they can see what is going on if housed. Let them have social contact with other cattle.
- Provide safety gates and escape routes in bull housing for stockmen.
- Bulls used for artificial insemination require special care, training and understanding.

Milking

- Milking must be a pleasant experience for a cow. Check the routine regularly and the equipment.
- Handling should start with the heifer as a calf.
- Herd size should be kept to about 100 cows.
- Animals should be run in similar age groups if possible.
- Cows should be dehorned.
- Feed and watering points must be adequate for the number of stock to avoid social conflict.

References

1. Bines, J.A. (1976) 'Regulation of food intake in dairy cows in relation to milk production.' *Livestock Prod. Sci.* **3**, 115–28.
2. Blackmore, D.K. and Newhook, J.C. (1982) 'EEG studies of stunning and slaughter of sheep and calves: Part III The duration of insensibility induced by electrical stunning in sheep and calves.' *Meat Sci.* **7**, 19–28.
3. Blockey, M.A. de B. (1981) 'Development of a serving capacity test for beef bulls'. *Appl. Anim. Ethol.* **7**, 307–19.
4. Brantas, G.C. (1968) 'Training, eliminative behaviour and resting behaviour of Friesian–Dutch cows in the cafetaria stable'. *Z. Tierz. Zuchtungsbiol.* **85**, 64–77.
5. Broom, D.M. (1978) 'The development of social behaviour in calves'. *Appl. Anim. Ethol.* **4**, 285 (abstr).
6. Broom, D.M. and Leaver, J.D. (1978) 'Effects of group-rearing or partial isolation on later social behaviour of calves'. *Anim. Behav.* **26**, 1255–63.
7. Curtis, S.E. (1981) 'Environmental management in animal agriculture'. *Animal Environment Services*: Mahomet, Illinois, pp. 1–382.
8. Darling, F.F. (1952) 'Social life in Ungulates. Structure et physiologie des sociétés animales'. *Int. Cent. Nat. Rech. Scient.* **34**, 221–6.
9. Dufty, J.H. (1981) 'The influence of various degrees of confinement and supervision on the incidence of dystocia and still births in Hereford heifers'. *N.Z. vet. J.* **29**, 44–8.
10. Edmunds, J. (1974) 'Multiple suckling of dairy cows'. *N.Z. J. Agric.* **129**(5), 56–62.

11. Forbes, J.M. (1970) 'The voluntary food intake of pregnant and lactating ruminants: A review'. *Brit. vet. J.* **126**(1), 1–11.

12. Gadbury, J. (1975) 'Some preliminary field observations on the order of entry of cows into herringbone parlours'. *Appl. Anim. Ethol.* **1**, 275–81.

13. Hafez, E.S.E. and Lineweaver, J.A. (1968) 'Suckling behaviour in naturally and artificially fed neonate calves'. *Z. Tierpsychol.* **25**, 187–98.

14. Hancock, J. (1950) 'Grazing habits of dairy cows in New Zealand'. *Emp. J. Exp. Agric.* **18**, 249–63.

15. Hancock, J. and McArthur. A.T.G. (1951) 'Tips on cow management arising from grazing behaviour studies'. *Proc. Ruakura Frmrs Conf.* pp. 32–8.

16. Hedren, A. (1971) 'The milk cow and the standing. Studies of the behaviour of milk cows in tying stalls'. *Rpt Agric. Coll. Sweden*, pp. 1–56.

17. Hoffman, M.P., and Self, H.L. (1973) 'Behavioral traits of feedlot steers in Iowa'. *J. Anim. Sci.* **37**(6), 1438–45.

18. Hudson, S.J., and Mullord, M.M. (1977) 'Investigations of maternal bonding in dairy cattle'. *Appl. Anim. Ethol.* **3**, 271–6.

19. Hurnik, J.F. (1973–4) 'Behavioural analysis of heat symptoms'. *Dairy Industry Res. Rept*, Ontario Agric. Coll. Uni. Guelph, pp. 79–85.

20. Hurnik, J.F., King, J.G. and Robertson, H.A. (1975) 'Estrous and related behaviour in post-partum Holstein cows'. *Appl. Anim. Ethol.* **2**, 55–68.

21. Kiley, M. (1976) 'Successful formula for multi-suckling'. *Brit. Fmr Stockbr.* **5**(119), 28.

22. Kilgour, R. and Campin, D.N. (1973) 'The behaviour of entire bulls of different ages at pasture'. *Proc. N.Z. Soc. Anim. Prod.* **33**, 125–38.

23. Kilgour, R., and Mullord, M. (1973) 'Transport of calves by road'. *N.Z. vet. J.* **21**, 7–10.

24. Kilgour, R., Skarsholt, B.H., Smith, J.F., Bremner, K.J. and Morrison, M.C.L. (1977) 'Observations on the behaviour and factors influencing the sexually-active group in cattle'. *Proc. N.Z. Soc. Anim. Prod.* **37**, 128–35.

25. Kilgour, R., Winfield, C.G., Bremner, K.J., Mullord, M., deLangen, H. and Hudson, S.J. (1975) 'Behaviour of early weaned calves in indoor individual cubicles and group pens'. *N.Z. vet. J.* **23**, 119–23.

26. Klopfer, F.D., Kilgour, R. and Matthews, L.R. (1981) 'Paired comparison analysis of palatabilities of twenty foods to dairy cows'. *Proc. N.Z. Soc. Anim. Prod.* **41**, 242–7.

27. Lalande, G., Beauchemin, K. and Fahmy, M.H. (1979) 'A note on the performance of Holstein Friesian veal calves raised to weaning individually or in groups'. *Ann. Zootech.* **28**(3), 235–8.

28. Macmillan, K.L. (1972) 'Detecting oestrus is the major problem'. *N.Z. Fmr* **93**(14), 14–15.

29. Macmillan, K.L. and Watson, J.D. (1971) 'Short estrous cycles in New Zealand dairy cattle'. *J. Dairy Sci.* **54**, 1526–9.

30. Marnane, N.J., Matthews, L.R., Kilgour, R. and Hawke, M. (1982) 'The prevention of bark-chewing of pine trees by cattle: The effectiveness of "repellents" '. *Proc. N.Z. Soc. Anim. Prod.* **42**, 61–3.

31. Marsh R. and Campling, R.C. (1970) 'Fouling of pastures by dung'. *Herb. Abstr.* **40**, 123–30.

32. Mattner, P.E., George, J.M. and Braden, A.W.H. (1974) 'Herd mating activity in cattle'. *J. Reprod. Fert.* **36**, 454–5.

33. Pain, B.F., Leaver, J.D. and Broom, D.M. (1974) 'Effects of cow slurry on

herbage production, intake by cattle and grazing behaviour'. *J. Brit. Grassld Soc.* **29**, 85–91.

34. Renger, H. (1975) 'Aggressives Verhalten von Bullen dem Menschen gegenuber'. *Diss. Uni. Munchen*, pp. 1–80.

35. Ruckebusch, Y. (1975) 'Feeding and sleep patterns of cows prior to and post parturition'. *Appl. Anim. Ethol.* **1**, 283–92.

36. Ruckebusch, Y. and Bueno, L. (1978) 'An analysis of ingestive behaviour and activity of cattle under field conditions'. *Appl. Anim. Ethol.* **4**, 301–13.

37. Sainsbury, D.W.B. and Sainsbury, P. (1979) *Livestock, Health and Housing*, 2nd edn., London: Bailliere Tindall, pp. 1–388.

38. Sambraus, H.H. (1980) 'Humane considerations in calf rearing'. *Anim. Regul. Stud.* **3**, 19–22.

39. Schake, L.M. and Riggs, J.K. (1969) 'Activities of lactating beef cows in confinement'. *J. Anim. Sci.* **28**, 568–72.

40. Schein, M.W. and Fohrman, M.H. (1955) 'Social dominance relationships in a herd of dairy cattle'. *Brit. J. Anim. Behav.* **3**, 45–55.

41. Schloeth, R. (1961) 'Das Sozialleben des Camargue-Rindes'. *Z. Tierpsychol.* **18**, 574–627.

42. Seabrook, M.F. (1972) 'A study to determine the influence of the herdsman's personality on milk yield'. *J. agric. Labour Sci.* **1**, 45–59.

43. Selman, I.E., McEwan, A.D. and Fisher, E.W. (1970) 'Studies on natural suckling in cattle during the first eight hours post partum. II Behavioural studies (calves)'. *Anim. Behav.* **18**, 284–9.

44. Stobbs, T.H. (1974) 'Components of grazing behaviour of dairy cows on some tropical and temperate pastures'. *Proc. Aust. Soc. Anim. Prod.* **10**, 299–302.

45. Stricklin, W.R., Graves, H.B. and Wilson, L.L. (1979) 'Some theoretical and observed relationships of fixed and portable spacing behaviour of animals'. *Appl. Anim. Ethol.* **5**, 201–14.

46. Trim, P.E. (1975) 'Human error involved in dairy cow submission to artificial breeding'. *Proc. N.Z. Soc. Anim. Prod.* **35**, 58–60.

47. Wagnon, K.A. (1963) 'Behaviour of beef cows on a California range'. *Calif. Agric. Exp. Stat.* Bull. No. 799, 57 pp.

48. Warnick, V.D., Arave, C.W. and Mickelsen, C.H. (1977) 'Effects of group, individual and isolated rearing of calves on weight gain and behaviour'. *J. Dairy Sci.* **60**, 947–53.

49. Webster, A.J.F. (1981) 'Optimal housing criteria for ruminants'. In *Environmental Aspects of Housing for Animal Production*, ed. J.A. Clark, London: Butterworths, pp. 217–32.

50. Willis, G.L. and Mein, G. (1982/83) 'Classical conditioning of milk ejection using a novel conditioned stimulus'. *Appl. Anim. Ethol.* **9**, 231–7.

51. Young, B.A. and Degen, A.A. (1981) 'Thermal influences on ruminants'. In *Environmental Aspects of Housing for Animal Production*, ed. J.A. Clark, London: Butterworths, pp. 167–80.

3 Sheep

Sheep have fulfilled man's need for meat, wool, milk since the dawn of history. They have even been used for sport (ram fighting) and wild sheep are still hunted.

Sheep vary in their distribution around the world. They are favoured in Asia Minor, Mediterranean countries, parts of Africa, southern Africa and Australasia. They are not popular in Europe and predator problems have greatly restricted their husbandry in North America.

Behavioural definition

In behavioural terms a sheep can be defined as a defenceless, vigilant, tight-flocking, visually-alert, wool-covered ruminant that evolved within a mountain grassland habitat. Its young (frequently multiple births) display a follower-type responsiveness. They suckle frequently and there is close bonding and strong imitation between young and old as they move within established ranges on well-defined tracks. Sheep show a seasonal pattern of breeding in temperate areas with a separate male subgroup structure at other times of the year.

The ram's arrival in the flock at mating time can stimulate the onset of breeding. Sheep react strongly by flight. As relatively defenceless animals they respond to disturbance by fleeing. They are visual animals and remain vigilant by paying visual attention to their mates and the environment. When disturbed, and at night they seek out higher ground. As flocking animals, they respond to disturbance and fear by forming tight groups, although within these flocks they still have an individual space requirement. Sheep are best handled in flocks.

Birth

Behaviour before birth
Near the time of birth, a ewe will try to find a lambing area which

is isolated from the rest of the flock. Studies have highlighted how certain areas in fields are more favoured for lambing than others. These areas are not necessarily well-sheltered.

Approaching lambing, ewes are more vigilant, more wary and graze less. The period of restlessness continues until she finally settles at a birth site. The ground is pawed frequently and with the onset of full labour and rupture of her waters, the precise position for lambing is established. From behavioural studies[5] in Australia and New Zealand, where these birth sites were marked by pegs, a number of conclusions can be drawn:

- The birth site is a focal point to which the ewe can return if she is disturbed.
- The site is clearly visible because of the paw marks made during labour, trampling of the pasture and mucus from the vulva.
- Once the lambs are born, there is transfer of attraction from the site to the newborn.
- The afterbirth is voided near the birth site. It is rarely eaten by sheep in contrast to cattle.

Such studies of lambing sites have highlighted these important implications for shepherding:

- Breeds differ in the average time they spend on the birth site, e.g. Australian Merino 2 hours and one strain of N.Z. Romney 6 hours.
- Young ewes such as 2-year-olds (two-tooths) that have not lambed as yearlings (hoggets) spend more time on the birth site than older ewes that have lambed more often.
- Ewes in labour can be seriously disturbed if moved daily onto a new pasture. This can occur in a drift or shedding (not housing) system. Problems arise because the ewes about to lamb may have selected a birth site and do not settle in the new field.
- If the ewe has to be caught and lambed, it is advisable to take her and the lamb back to the birth site if it can be found. Otherwise tying her to a peg will create an artificial birth site, preventing her from wandering away until she has accepted the lamb.
- Ewes and lambs should not be disturbed from the birth site until they are ready to move themselves.
- Spot-marking lambs with a raddle as soon as possible after birth will help avoid later recording errors. For example twins can be given the same colour code on the same place on the body. The

ewe and lambs can then be left until well-bonded before any further disturbance.

- Unless the ewes are well used to the shepherd before lambing, disturbance due to shepherding can result. Likewise poorly controlled dogs can be very disruptive.
- The shepherd should decide on either a very intensive system or an 'easy-care' lambing system where ewes are left to themselves to lamb. Most intermediate management systems between these can be disruptive to ewe–lamb bonding and in fact save few lambs.

The functions of the birth site are summarised in fig. 12. A basic rule of shepherding ewes lambing outdoors is to respect the birth site. This site can provide some valuable information for the shepherd. For example the afterbirth can be found and examined to see how many separate 'cords' there are on it. Each lamb has two veins and two arteries so actual number of lambs born to the ewe can be determined. This may not agree with the number of lambs the ewe has at foot.

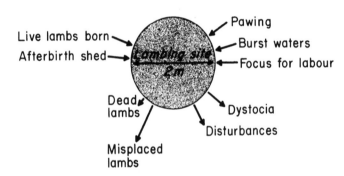

Figure 12 Birth site functions derived from ewe behaviour at lambing[19]

Ewe behaviour during birth

The ability to delay contractions (as seen in the cow) is not well developed in the sheep. In sheep, contractions are so rapid that partogram observations used 5-minute periods and most normal labours take less than an hour. Disturbance by the shepherd only interrupts the ewe's labour for up to 15 minutes.

The ewe's response to interruption can be modified to varying degrees by previous conditioning. Thus it is good practice to accustom a mob of lambing ewes to the shepherd, dogs, motorbikes, vehicles and so on in the period leading up to lambing.

In detailed studies[5] of over 300 observed births, with normal presentation of head and forelimbs first, labour took about 1 hour. The birth interval between first- and second-born lambs in twins was less than 1 hour in 95% of cases. Once the shoulders are born, the lamb tends to slip out when the ewe stands up. The lamb will be dropped within the confines of the lambing site, and if not disturbed the ewe stays on this area where she will bond with the lamb or lambs dropped there.

After rising, the ewe chews at the covering membranes and licks the fluid from the lamb, often starting from the head. Ewes lambing for the first time need to show this important critical response of immediately turning completely about after delivery and attending to the lamb.

The lamb being precocial soon sits upright, shakes its head and moves about while being licked by the ewe. As the lamb tries to rise to its feet the ewe's licking may knock it over. After the lamb can stand, it is normal for the ewe to start to pay attention to the perianal region and this helps the lamb orient itself towards the udder area. Ewes with strong maternal responses arch the back which tilts out the teat slightly to help the lamb locate it.

Birth difficulties

There are a number of birth malpresentations possible in ewes with single lambs (e.g. head back, one or both forelegs back, breech presentation and so on) and these can be further complicated where the ewe gives birth to more than one lamb (e.g. two legs present but belonging to different lambs). Added to this is the condition described as dystocia where the lamb is too large for the birth canal, or it sticks in the canal during delivery. Often birth is delayed at the point where the lamb's head is out and it then swells up.

If labour has been long and difficult, the ewe often moves off the lambing site and finally ignores the lamb once it is born. Observations show that after about 1 hour in labour, the ewe becomes restless and will leave the birth site, bearing down and 'pressing' as she moves around. When the lamb is finally born, the ewe has lost the smell orientation of the birth position which helps her locate lambs. Thus ewes with dystocia inevitably are poor mothers and their lambs suffer high mortality.

In twins where the second-born lamb suffers dystocia, the first-born lamb may have wandered off following other passing ewes or may be prone to being stolen by other ewes. Then it can be left to die of starvation-exposure.

In cases of difficult birth and prolonged labour the ewe's birth fluids become dried out and this makes the delivery of the lamb more difficult, especially if it has a thick birth coat.

Many studies have described the nutritional management of the ewe flock during the last stages of pregnancy to produce lambs of optimal birth weight for survival and to avoid dystocia. However, the optimal combination of management factors is often difficult to achieve regularly. This is complicated further by not knowing whether the ewe is carrying singles or multiples.

Ewe–lamb bonding and recognition

Experimental work where lambs were removed at birth and then returned to the ewe at various time intervals showed that there is a critical period of about 4 hours after birth after which the ewe will treat its returned lamb as a stranger. Also cases have been documented of lambs joining a ewe that is licking her own newly-born lamb a few hours after birth and being accepted readily.

Once this critical period is over, alien lambs will be severely butted and rejected. The ewe–lamb bond is established quite soon after birth and is heavily dependent on smell and taste. Research[3,4] has established that it is the odour from wool and perianal secretions which allows identification and not the smell of the ewe's own milk on the lamb or the lamb's dung and urine.

The vocal calls made by sheep and lambs are generalised location calls and do not finally identify either the ewe or the lamb. The recognition of the ewe's own lamb at close quarters is made by smell although calling helps them locate each other at greater distances. When ewes and lambs were released at either end of a 13 m run, on meeting, the ewes sniffed the lamb's flank and perianal region for recognition, and only after positive identification did the ewe allow the lamb to suckle[2].

Lamb stealing

This is a fairly common occurrence in sheep, especially in ewes with strong maternal motivation that are within a few hours or days of lambing. These ewes lick the new-born lambs of other ewes, allow them to suckle and will take the lamb from its correct mother. Ewes that steal lambs days ahead of lambing can repeatedly disturb the bonding process between other ewes and their lambs. Finally when they give birth to their own lambs they often make a poor job of rearing them.

Lamb stealing is made worse where ewes are lambed in groups

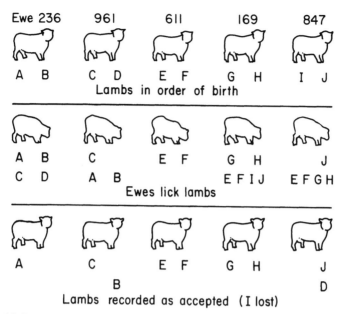

Figure 13 Documented observations of exchanged new-born lambs among five ewes lambing close together

at high concentrations and where they cannot escape from the rest of the flock to lamb unmolested.

Lamb stealing can cause considerable complexity and error in recording the correct parentage of lambs as shown by a documented case in fig. 13.

In this group of five ewes giving birth to twins, a number of ewes licked their own and other nearby lambs. Some lambs were interchanged, some accepted and some lost. The shepherd's observations led him to make 40% errors in linking the wrong ewe with the wrong lamb or recording a lamb as being born as a single when it was born as a twin. In a range of studies in many different topographical areas with several breeds, lamb stealing ranged from 6–18%.

Suggestions to prevent lamb stealing are:

- Identify the ewes that persistently steal lambs and remove them from the lambing mob or restrain them. These lamb stealers are usually 1–4 days from lambing.
- Restrict general ewe mobility in the lambing area by central zig-zag temporary fencing.
- Spread the ewes out over a wide area and if possible lamb on broken, undulating terrain.
- Form flocks of ewes with a spread of lambing dates.

It has been suggested by some farmers that access to salt or other mineral licks will reduce the incidence of lamb stealing but this has not been put to properly controlled experimental tests.

Fostering orphan lambs

Many techniques are used to foster orphan lambs onto other ewes. Some of these are:

- Give the ewe the orphan lamb while she is still in labour with her own lamb, covering the orphan in the ewe's birth fluids.
- Skin a ewe's own dead lamb and tie the skin on the orphan lamb until it is accepted.
- Restrain the ewe tightly in a pen to prevent her butting until she accepts the orphan.
- Restrain the ewe to a peg, tree or fence in the field until she accepts the orphan. This is best done where she lambed.
- Wash the orphan lamb with a detergent to remove all alien smells and foster it on to a restrained ewe.
- Treat the lamb's back with a strong smelling liquid such as turpentine, kerosene, whisky, deodorant, ladies' scent or a proprietary aerosol spray of anaesthetic and odour. Insert some of the same liquid up the ewe's nostrils.
- Tranquillise the ewe using drugs. When she recovers she may accept the lamb.
- Express some of the ewe's milk over the lamb's back and leave in a confined space. This is not guaranteed to work.
- Tie a dog up within sight of the ewe to encourage her to protect and mother the orphan lamb.
- Tie an orphan lamb to the ewe's own lamb by means of a leg hobble so that they cannot be separated and the ewe has to accept both.
- By means of an electronic dog collar, give the ewe a reduced shock every time she butts the lamb. This has not been successful in practice.

Experience shows that the chances of successful fostering are low if the ewe has had a prolonged and difficult labour or has a poor milk supply. Additional lambs fostered on to ewes should be the same size as the ewe's own lambs. Adoption up to 4 hours after birth is relatively simple but after that time the ewe's sense of smell must be overriden. Few good trials have been carried out to compare all the fostering methods available.

Some orphan lambs survive by suckling from any ewe in the flock through her back legs rather than by the normal position alongside the ewe facing her rear (parallel inverse position). Suckling from behind they cannot be seen or smelled so easily by the ewe. These lambs are easily recognised by accumulation of faeces on their heads. They have also learned to suckle when the ewe's own lamb is suckling.

Lamb's response to the ewe

Once the lamb gets to its feet, it works its way along the flank of the ewe seeking the teat under either the forelegs or the hind legs. They are assisted in the correct direction by the ewe licking the perianal area. If no ewe is present, they will pursue the same searching action along walls and fences.

Research[6] that used ewes with covered udders showed that the lamb's suckling drive reaches a peak about 3 hours after birth and then declines to a very low level of effort after 12 hours. Lamb survival depends on finding the teat as early as possible after birth.

New-born lambs within a few minutes of birth are attracted to moving objects such as other ewes, shepherds, dogs and vehicles. This can inadvertently attract them away from their birth site and increase the risk of mortality. Within 3 days of birth, the lamb's tendency to follow moving objects becomes fixed on its own mother, and a few days after that, the lambs will move if other ewes are moved showing the very early development of the flocking and the 'follower-type' reactions. It is through this follower behaviour that the lamb learns a great deal about its environment, e.g. the sheep tracks and territory on open grazing lands.

Lambs 'heel', stay with their dams and suckle about once every hour during daylight. Their nocturnal habits have not been documented. After the first day or so when called by the ewe's bleat, the lamb travels to the ewe rather than the ewe to the lamb.

Ewe's response to multiple-born lambs

The care of multiples generally causes the ewe extra maternal problems. However if multiples are born with ease with a short time interval between each birth, and they are dropped on the same birth site, then an effective dam—offspring bond soon occurs. The lambs also bond well with each other and thereafter stay and move together.

Research[5] showed that breeds differed in their response to multiples with the Australian Merino having a greater tendency to leave one lamb behind.

Some components of the complex trait called maternal behaviour have not yet been adequately researched.

Lamb survival – ewe's fault or lamb's fault?

Sheep breeders in their selection programmes often have to apportion blame for lamb mortality to the dam or the lamb. This is often very difficult to do.

Studies[18] on steep hill country showed that after being born, lambs slipped from the birth site while the ewe remained on the site drawn by the smell of the birth fluids. Lambs were observed to slip 30 m down the hill and suffer physical injury. Licking the lamb by the ewe may push it down the hill and in the case of multiples, lambs may slide in different directions or become separated. Lambs roll through fences, into swamps, holes and so on. The danger of slip is greatest in the first 3 hours after birth. Thereafter most lambs can follow ewes up quite steep slopes.

If the ewe lambs in a safe, sheltered, level site then there is less chance of blaming her if the lambs die. A common feature of some ewes is that although they are attentive and lick the new-born lamb, they do not stand still and orient themselves correctly to let the lamb suckle. Here a lamb's death could be attributed to the ewe.

Ewes may cause the lamb's death by chewing the navel or tail causing haemorrhage or stand on the lamb and kill it. But the area of major concern is with the ewe which accepts and mothers one lamb and ignores the others until they die of starvation-exposure. More study is warranted in this area because of the range of factors involved in mortality (see fig. 14).

Thus where decisions have to be made in apportioning blame for lamb mortality, it is better to use a classification of 'don't know' to code the lamb's death rather than guess.

Shelter has been demonstrated[20] to be very important in improving lamb survival and pre-lamb shearing is used in some countries to encourage the ewe to seek shelter and hence lamb in these positions.

Pen lambing

There are many systems of sheep production where the ewes are placed in individual pens either in the lambing field or indoors. The reasons are these:

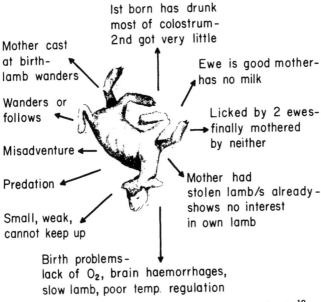

Figure 14 Summary of factors leading to lamb deaths[19]

- It provides shelter and hence improves lamb survival.
- It ensures dam–lamb bonding which again improves survival.
- It ensures correct pedigreeing by avoiding lamb stealing and lamb swapping. This is especially important with large litters of lambs.
- It allows easy surveillance by the shepherd.
- It provides protection from predators.

There are some interesting behaviour aspects of pen lambing that as yet have not been researched. Experience in Australia with the highly fertile Booroola Merino shows that the effect of lambing in pens has been to reduce the strength of the bond. The small pens kept the ewe and the lambs together. When the lambs went outdoors 24 hours later, however, the family group did not keep together.

Lambing in pens means the ewes have to be handled more and there may be problems of hygiene as lambing progresses. For these systems to work well, more information is needed about the signs of imminent parturition, the time the ewe needs to settle into the new environment and time needed to establish a pen birth site.

Identifying lambs and their dams

Where detailed recording of lamb-to-dam cannot be done at birth, lambs can be identified to their dams by marking the ewe's udder with coloured chalk or raddle. As the lamb sucks and bunts the udder, its head is covered in raddle.

In practice this marked-udder method or MUM allows the splitting of a large flock of ewes and lambs into smaller groups to suit feed supply. It is also useful in sorting out mobs that have been accidentally mixed, and in removing individual ewes (e.g. with disease) along with their lambs from a larger group.

However, as there are a limited number of colours of raddle, identification can only be done on small groups at any one time. There is also the chance (which can be as high as 17%) that the ewe suckling the lamb was not born to her. This chance of error can be high with large litters of lambs born at pasture.

Rearing

With ewes and lambs grazed at pasture, once a good dam–lamb bond has been established then there are few behavioural problems during the rearing stage up to weaning. The suckling of the lamb is dictated greatly by the ewe's milk supply and the lamb may be weaned long before the offical weaning by the farmer.

The weaned lamb will continue to follow its mother after weaning for a very long period if allowed. This follower relationship can last for several years and is seen in feral sheep and in sheep that graze open range.

In intensive rearing systems where lambs are removed from the ewe a few days or a few weeks after birth, some interesting behavioural implications can arise. Lambs can be fed from individual bottles and teats but where large numbers are involved, rows of rubber teats and bulk-milk containers are used. Setting the teats below a ledge covered in wool is used to assist the lambs to learn where the teats are in the early stages. Each teat is generally separated by a divider to stop lambs chewing all the teats along the row. Lamb feeding becomes a social event. In these pens 8 teats have been used successfully to feed 16 lambs, in an *ad libitum* cold milk feeder system.

Disturbance of lambs in indoor, intensive rearing systems will generally result in their going to the teats to drink as happens in a flock at pasture. Growth rates of lambs near doorways or other points of disturbance will be greater than those less frequently disturbed in quiet corners of the shed.

In some studies of pen rearing of lambs navel sucking, ear sucking and other so-called vices have been reported. These activities parallel those shown in bucket-reared calves and require a smaller teat orifice to force lambs to suckle longer for the available milk.

Lambs quickly learn to creep feed whether at pasture or indoors. Finding the entrance to creeps or to holes in fences or walls seem to be aided by the lambs' play behaviour. Lambs at pasture start to nibble grass from a few weeks and are effective ruminants after about 3–4 weeks of age. Grazing can be well established by 6 weeks.

Routine management

This includes the following:

- Ear marking – snipping bits from the ears to denote ownership, sex and age.
- Ear tagging with various tags using a wide range of insertion methods.
- Tattooing numbers in the ear.
- Docking (tail removal) by knife, rubber rings or cauterising iron.
- Castration by knife, rubber rings or pliers that crush the spermatic cord.
- Mulesing – cutting off the skin wrinkles around the crutch and tail of some Merinos.
- De-horning in some breeds.

Lambs tend to be slightly shocked after such handling and they must be given time to mother up with their dams before being returned to bigger fields or range. Greater emphasis on labour saving has seen the recent development of many devices like chutes for restraining animals during some of these management routines.

Reproduction

Sheep have a specific breeding season and most of the between-sex mating responses fall during this period. Breeding is triggered by a number of responses, the most important of which is a decreasing daylight pattern in temperate regions. In the tropics sheep breed all the year.

The ram

Rams reach sexual maturity between 6 and 9 months of age and can be regularly used as yearlings.

Studies of wild and feral sheep show that within the male group outside the breeding season, a great deal of sexually related behaviour takes place. This can be seen in groups of rams on farms. They spend time butting, threatening, head-to-head fighting and more dominant rams will mount other males. Masturbation also occurs. The seasonal variation of this activity has not been fully examined.

Research[23] has shown that all-male rearing up to weaning (4 months old) may focus the male's attention on other males and these mono- or homosexual responses can continue into puberty. Rearing rams in isolation from females up to puberty did not reduce their interest in females after puberty. Interest in females took only a few minutes to develop. This contrasts with the response of pre-pubertal ram lambs to females when they move towards the ewe, stand passively beside them, sniff at their wool and bleat strongly.

Figure 15 shows a flow chart of sequence of responses during the courtship process between ewe and ram starting with the naso—ano sniff-hunt. Once a ewe is found by sniffing this leads on to either the flehmen or a range of courtship responses including

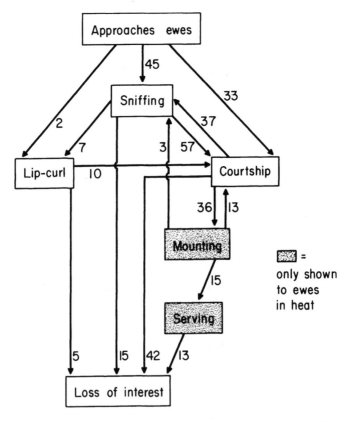

Figure 15 Sequence of ram courtship responses to oestrous ewes. Numbers are the number of times each behaviour occurred in a sample of twenty ewes (after Ross, K.R. (1973) *MSc. Thesis*, Massey Univ., N.Z.)

nudging and pawing the ewe, various vocalisations, intention mounts or even some butting. The ewe's responses of urinating, looking back at the ram, tail-fanning after being nudged and generally standing still, further stimulates the ram to mount. After the ejaculatory thrust the ram slowly dismounts, stands with head lowered for a period (latency period) before he recovers to begin again.

After finding the ewe is not in oestrus by the sniff-hunt process, or after effective mating, the ram moves on to search for other ewes. He may return later in the day to repeat-serve the ewe if still in oestrus[13].

Ram—ewe mating ratio

Traditionally a mature ram has been joined with 40–50 ewes. Research[7] has shown that a ram can effectively mate many more ewes without affecting lamb production and one ram per 100 ewes is now normal practice in New Zealand. Certain highly selected rams can be joined with as many as 200–400 ewes with successful results. Observation of the ram's rate of recovery after ejaculation and the range of responses to many different ewes would be the most useful data to determine a mating ratio. The distances walked by a ram each day can be considerable. One ram tracked at the start of the breeding season travelled 12 km in a day while sniff-hunting ewes (see fig. 16).

Rams must be physically fit and some farmers pre-condition their rams by grazing them with the ewes before the breeding season begins. Many rams are grossly overfat and out of condition before mating, especially for hill country conditions.

Some rams when joined with ewes in range conditions or in large fields attract harems and spend their time with these small groups which remain separate from the main flock. Certain breeds such as the Merino seem to be more prone to this than British breeds. Regular gathering (mustering) at mating is thus needed to rejoin these rams with the main group.

Sex drive (libido)

In the interest of farmers, being able to predict a ram's sexual drive (libido) before mating would be a help in establishing suitable mating ratios. The first problems arise with very active rams which mate well but are infertile. Valuable time may be lost before this is noticed by ewes returning to oestrus. Semen tests can be carried out before joining but these too have limitations.

The other problem is 'inhibited' rams or those that will not work

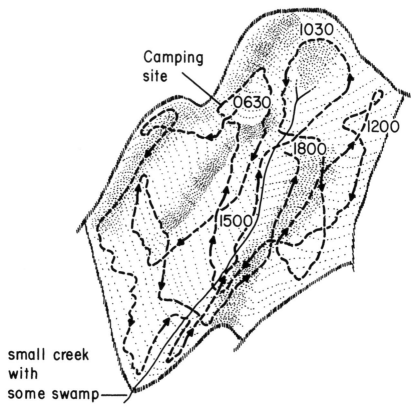

Figure 16 Track made by ram on steep hill country, seeking ewes in oestrus (after Kilgour, R. and Winfield, C.G. (1974). *N.Z. Jl Agric*, **128**(4), 17–23)

when joined with ewes and the reasons for this are not fully known. In practice, it seems safe to accept that if a ram has started working, he will continue to do so unless ill-health or some calamity stops him. Some rams are slow to start working but once started have no problems.

A test has been developed for research purposes to measure sexual drive and dexterity in the ram. It is expressed as the ratio of mounts to services (ejaculations) when given 4–5 oestrous ewes in a small pen for 20 minutes. Rams scoring well on this index have performed well in field situations. As the test needs to be carried out about three times in the weeks up to mating, it would seem to be rather impractical for farmers and more work is needed to develop an easier, shorter test.

The normal range from rams studies is one ejaculation every 4 to 5 mounts. One poor performer had one ejaculation per 80 mounts. Efficient rams given many ewes need to ejaculate after very few mounts.

A point to remember is that although libido tests on young rams may indicate their behaviour (sex drive), they are not necessarily accurate indicators of the ram's physiological development. Libido tests are better for older rams which are sure to be fully developed. Puberty in both sexes requires both a physiological and behavioural maturing of the animal.

Social dominance in rams

In group mating situations, rams show a social dominance structure often carried over from their pre-joining social status. Observers[8] have shown that subordinate rams did a lot of hunt-sniffing and mated ewes in the early and late oestrus with dominant rams mating during peak oestrus. However, other observations have not been able to relate social ranking with field-mating behaviour. Social dominance will only be of concern where there are too many rams for the ewes in oestrus and they waste time and effort fighting. In very small fields dominant rams are able to prevent subordinate rams mating. The dominant ram does most of the mating of ewes in peak oestrus. Lambing percentage can be severely affected if a dominant ram in a group is infertile.

The older male and teaser rams

Experience in older rams can make up for many disabilities. The older rams usually have good knowledge of the terrain and with growing dominance status may claim important territory in shade or near water and find oestrous ewes when they come close. With older rams there is a greater risk of health problems preventing them from mating such as footrot and poor physical condition. Their management needs attention at all times of the year, not just at mating.

The rams that are vasectomised and made into teasers are generally animals culled for various reasons. Mid-dexterity animals make good teasers as in their frequent mounts they leave good crayon marks on the ewe. Teasers need to be in as good physical condition as entire rams otherwise their sex drive quickly wanes.

Wethers and ewes treated with hormones have both been used successfully as teasers in research work but there is only limited experience of their use on farms.

The female

Puberty in the ewe lamb appears to depend more on weight than on age in the British breeds. Most will reach puberty at 8–10 months of age but 35 kg seems to be a minimum live weight for the expression of oestrus. Ewe lambs start their breeding season about 4–6 weeks later than mature ewes.

The sheep's oestrous cycle varies from 16 to 20 days with oestrus lasting from 6 to 16 hours. In most temperate breeds, the breeding season lasts for around 3 months. However, most farmers want their ewes to become pregnant in the first 2–3 cycles of the season to restrict the lamb drop for management reasons.

Ewe lambs do not show the ram-seeking behaviour found in some older ewes and they also have a reduced level of courtship response, stimulating the ram less than older ewes.

The behavioural responses of most older ewes which stimulate the ram to further activity are tail-fanning, looking back, scrotal nuzzling of the ram and standing very still. These responses improve with age and experience. The most critical response for male mounting however is immobility. Movement of the ewe signals absence of oestrus.

Tail-fanning appears to be a response to assist the entry of the ram's penis by side flexion of the tail. This is of practical importance in breeds with long woolly tails and fat-tailed sheep have special mating problems. Where tails are docked no problems arise. The wool is usually shorn from long-tailed sheep before mating to help the ram.

Ewes given hormones (oestrogens) to increase the level of oestrus or to synchronise oestrus show a greater tendency to seek out the ram, fight with other ewes around the ram and butt him when he mounts other ewes. However, the use of antioestrogens to increase ovulation rate may also have a deleterious effect and suppress the oestrous behaviour of the ewe. More practical research is needed in this area.

Rams stimulate the onset of the breeding season and oestrus in the ewe. There appears to be variation between rams within breeds and between breeds in the amount of pheromone (male stimulant) located in the grease exuded from the ram's skin and fleece. The pattern of these secretions is also seasonal.

Concern is often expressed by farmers that poor flock fertility may be due to practices such as mustering, yarding with dogs, shearing, drenching and dipping. There is a little evidence that the stress of shearing, reducing the sheep's temperature and increasing its appetite, does lower conception. Physical handling seems of very little significance.

Mating systems – general points

Pen mating

In pen mating or hand mating systems, groups of rams are held at some central location and ewes in oestrus (identified by harnessed teasers) are brought to them to be mated singly or in small groups. Likely problem areas to avoid as seen in practical studies were:

- If a peak oestrous ewe was in a pen with a ewe in declining oestrus, the ram mated the former ewe first while the latter ewe went off heat further and finally refused to stand for the ram's advances.
- Older ewes were mated before younger ewes when in the same pen.
- Rams mated ewes of their same breed before ewes of other breeds when penned together.
- Feeding the rams upsets their courtship and mating routine. Feeding takes priority over mating. Changing feeds may scour (cause diarrhoea) the rams which in turn can reduce mating activity.
- The presence of other rams in close proximity to the working rams caused disturbance. This 'audience effect' stimulated the mating rams to fight the watching rams.
- Keep farm dogs away from sheep mating in pens.
- Illegible crayon marks made by the teasers are always a problem as it is hard to discern whether they denote successful mating or rape by the ram. Careful fitting of the harness so that it fits well and does not chafe the ram is important and a good sequence for the coloured crayons is yellow, green, red, blue, purple.

Artificial insemination (AI)

With artificial insemination, the main problems arise with training rams for semen collection. Some points to achieve good results are:

- Get the rams used to the environment, such as the pens, races, gateways for a week or so before collection.
- Hand-feed the rams to assist taming and getting them used to their handlers.
- Patience is required at all stages.
- Rams may be reluctant to work in high temperatures and out of season.

The weakest point in the whole AI system is the rough handling ewes may receive before and during insemination. A good system must not stress ewes during insemination, and an efficient routine is essential.

Mating in the tropics

Ewes in the tropics return to oestrus some weeks after lambing and can be mated by rams that will work all year. However, there may be intense competition between rams for the few ewes on heat at any one time, and it may be difficult to find foster mothers for orphan lambs. Sheep in the tropics require different management.

General management

Flocking

The best management of sheep will take into account their social behaviour. Where small groups of sheep were put through a maze, groups of two or three animals regularly split up but four sheep acted together and quickly learned their way out. Above seven sheep, they split into groups generally of about four. Dog trials in New Zealand use three sheep as the most challenging group size to handle.

Clearly data from less than four sheep cannot be accurately applied directly to larger groups. In small flocks sheep are able to keep eye contact with at least one other animal. This enables them to flock quickly if disturbed. In large open territory, this small group behaviour is very obvious from aerial photography studies. The night camping habits of sheep groups on high ground are well recognised and can concentrate soil fertility through the build-up of dung and urine at these points. Weeds may also increase in these areas.

Grazing

Sheep undertake periods of grazing that are slightly less rigid than cattle but are clearly diurnal and change as the season dictate. Up to 8 hours can be spent grazing each day in temperate climates but even when feed is in short supply and they are hungry, they will not increase their grazing time much above 10 hours.

Sheep graze within a 'home range' territory on open grazings in Britain and semi-arid areas; their diet is strongly controlled by the amount of food near the watering point. They run out tracks from these points like spokes in a wheel and then graze on subtracks within the main spokes. This range movement pattern based on a water hole is called a piosphere. Kilgour[16] suggested that a raceway directing the sheep beyond the overgrazed areas around the water holes would allow better grazing control. The raceway could be constructed using either solar- or wind-powered electric fences. Other funnel divisions could be added later for even better grazing control. This is shown in fig. 17. As the distance the stock graze away from the water hole is limited to how far they have to return for a drink, the funnel fences need not extend beyond this point.

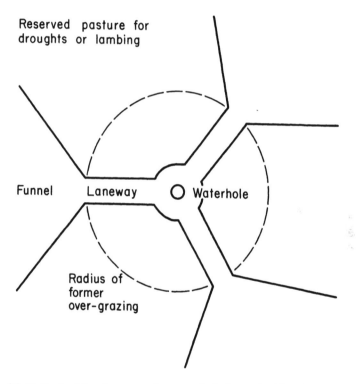

Figure 17 Method of fencing around a water hole in semi-arid region, recognising sheep and cattle behaviour[16]

In hill country sheep develop a system of tracks along hillsides like contour lines on a map. This is because they prefer to eat reaching up above them, rather than below them as they walk along the hillside. However, when hills become extremely steep (more than about 45°) sheep form paths straight up and down the hill face and graze on either side as they walk upwards.

The placing of fences should take into account these sheep tracks if effective grazing is wanted. Badly placed fences or gates in relation to the sheep's movement patterns in the grazing area can lead to injury and smothering during mustering on hill country. The tracks become part of the web of movement for a 'follower' type species like the sheep and their close study would be justified. On flat areas the characteristic wavy pattern of the tracks is due to the sheep's need to watch behind at regular intervals as they walk along. This is because of the blind spot behind the head.

Farmers can study sheep movement within localised areas by marking individual animals with bright coloured raddle. This can provide information to help decide where to put fences, gateways, water troughs and supplementary feed blocks.

Feeding problems

Sheep develop fixed eating patterns although their diet can be very variable. Serious problems arise when this pattern has to be abruptly changed for example in drought conditions, floods, heavy snowfalls or where they have to be transported long distances by boat. Some sheep will not accept a diet change and die of starvation.

The following points should be considered:

- Pre-condition the animals early in life to the feeds or supplements they may be expected to eat in later emergencies. Research[15] has shown that sheep have good long-term memories. Lambs learn quickly from their mothers to eat grains and other new foods before weaning and this should be exploited.
- Introduce the new supplement gradually into the current diet. If necessary add it into a mixture with known acceptable food. This is not always possible in disasters.
- Use older experienced sheep to teach younger ones to eat new feeds.
- Find out which sheep are eating the supplement successfully and which are not. This can be done by using a feeder into which the sheep must put their heads and as a result press against a piece of sponge with coloured dye in it. The non-coloured animals that will not eat can be removed and given other feed if available. In a training programme, the quick learners can be removed to allow more time and less competition for the slow learners. Problems can often arise with supplements fed to animals. The farmer intends them to be 'supplements' to the basic diet, but the animal starts to use them as 'substitutes'. An example is

where sheep may wait all day for the silage supplement or concentrates and not graze the pasture. The solution seems to be to feed the animals at times and in places in an irregular way so that they do not develop 'expectancies' or routine habits.

Controlling sheep predators

A small but significant percentage of lambs can be taken by predators in some countries. Coyote losses in Kansas are considered to be about 3% per annum; feral dogs also cause damage. Coyotes or domestic dogs worry mainly weak sheep at night and in general if the offending animals can be destroyed the problem is overcome. Most worried animals suffer throat or head injuries, though feeding starts with intestinal fat, hindquarters or ribs. Methods of reducing predation include:

- Locking stock up for the night. This practice is the most ancient.
- Removal of dead carcasses or other carrion which attracts predators. Carrion is too erratic for predators to live on alone.
- Fencing. Most canine predators go over or through fences and wire fences with or without electrification have been designed. Other big predators do break through these fences allowing canines in at these times. Fences are also costly and electric fences can be shorted by growing plants. Dogs can jump fences.
- Various frightening devices have been designed, e.g. lights, explosive devices, bells on sheep necks, radios or scarecrows. Predators soon get accustomed to these so periodic use only can be made of them.
- Guard men or dogs. Recent programmes to 'imprint' Komondor dogs as pups onto sheep have met with some success. These dogs keep other predators away from 'their' sheep. Training is costly and the result not always successful.
- Repellents. Dog collars with marking colours to indicate offending predators or poisons to kill the biting dogs have been used. Those that live become more wary. Hundreds of chemicals have been tried but none has completely succeeded in practice.
- Trapping, shooting. The effects are usually short-term as territories which are emptied are usually filled again very quickly. Some welfare concern has been expressed about trapping. Culling coyotes by shooting has helped reduce populations of younger pups leaving the den.

A combination of good fences, disposal of carrion and removal of offending animals are the key points in any predator control programme.

Handling facilities

Many different sheep handling systems and designs are used throughout the world and there is a constant need to find improvements to reduce labour and stress on the animals. Some basic principles were first outlined by an Australian sheep farmer called Hopkins. These are presented below with some modification based on other experience and research[14]:

- Oncoming sheep should not be able to see the operator.
- Keep the way ahead clear without blind ends. Studies show that sheep stop about 3 to 5 m from a dead end and retrace their steps — the front ones turning around and blocking the flow. Gates which lift in front of sheep and fall behind them are best. Drafting races are best close-boarded but at other places in yards, flow benefits by sheep being able to see other sheep.
- Sheep which are following should be able to see moving sheep ahead. In single races sheep disappearing quickly around a corner or into a dip have a pulling effect on other sheep behind.
- Sheep prefer to move in a straight line. They travel more freely in a wide race rather than a narrow one. Decoy sheep can be penned at awkward points to attract others. Mirrors are often used to draw sheep along races.
- Advancing sheep should not be able to see the sheep behind or they will stop and turn around.
- Sheep move best on level ground, but if this is not possible it is also recognised that they flow better up slopes than down them.
- Type of ramp floor seems unimportant as long as they have a good foothold. Sheep will readily climb quite steep steps into catching pens.
- Sheep move best from dark into light areas and generally dislike changes in light contrast. Areas where sheep move regularly should be well lit, e.g. raceways, woolshed catching pens and so on. The flow is heavily dependent on the lead sheep.
- Sheep do not like to see reflections or light shining up through slats, grating or through holes. Gratings must be laid across the flow path of the sheep, especially where they enter a new area.
- A lone sheep is inevitably agitated and stressed and this can be reduced by providing at least one companion. This applies to all situations including slaughter.

- Handling facilities should be examined from the height of the sheep rather than the human viewpoint to detect flaws in the design. Simplicity in design should be a major concern. Avoid the faults shown in fig. 18.
- Dogs. On range, dogs appear as predators to sheep and cause them to flock. In yards where contact is closer, confrontations are more direct. Dogs not needed should be tied up away from yards. If packs of dogs are needed to get sheep to flow through yards there must be major faults in design.
- Always question and carefully examine the latest fashion in yard design or equipment before spending money on it.
- Noise such as rattles is effective in moving sheep but should not be overused or it will lose its effectiveness when the sheep become used to it.
- Sheep do not like walking into their own shadow.

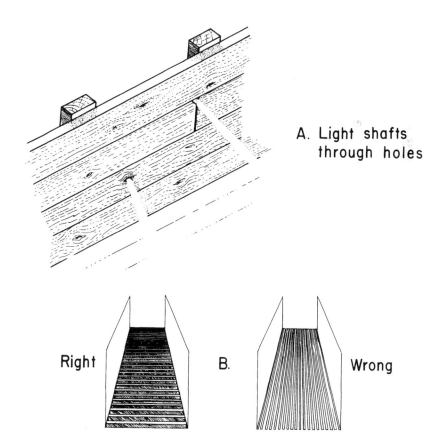

Figure 18 Examples of faults in sheep yard design which restrict animal flow (after Kilgour, R. (1977). *N.Z. Fmr,* **98**(6), 29–31)

There are many other aspects such as the effect of the sun's position, wind force and direction and noise that are claimed to affect handling. These need correct behavioural evaluation as does the value of a wide range of handling aids regularly appearing on the market.

Dipping

Sheep do not like going into water. As dipping is a stressful experience for sheep, the dip (whether swim or spray) should be sited at the side or away from the main handling area. Decoy animals can be used to draw sheep near the dip and then using surprise as with a falling trapdoor, they can be plunged into the dip. The smell of the dip however, is enough to make sheep unwilling to go along the race towards the bath.

With spray dips, curved races and wide entrances to the spray pen are a great help. Again decoy sheep can be used.

The same principles apply to facilities for washing sheep in plain water prior to slaughter.

Shearing

A good shearer can remove the fleece from a sheep in under 3 minutes. The stress of actual shearing is thus much less than the effect of losing the fleece, especially if the weather is cold. Shearing is a major physiological shock to a sheep and it responds by increased appetite. Extra feed, and shelter must be available to meet these needs.

Shearing using hand shears (blades) leaves much more wool on the sheep and this is still used where snowfalls are possible. Snow combs which leave more wool on the sheep when machine shorn have not been popular with shearers as they increase shearing time for each animal.

Chemical shearing where sheep are given drugs to remove all their wool has not been successful. Sheep have to have covers fitted before wool removal, it takes as long to pull the wool off as to shear it, and the sheep must be recovered after de-woolling to prevent cold shock and sunburn. It is too expensive.

The future will see much more automation of the shearing process where sheep are handled by machines rather than workers. A successful system is now in operation in Western Australia. The sheep's welfare will need to be closely watched during these developments.

Catching and lifting sheep

Accident statistics in New Zealand show that many farmers and shearers sustain back injuries before age 35. These happen when sheep are being caught and lifted for foot trimming, dagging, shearing and so on. The following points should be considered to prevent such problems:

- In small pens approach the sheep exploiting the blind spot behind its head when its head is raised.
- Place one hand on its muzzle and turn its head back along its side to face the rear.
- The sheep will collapse slowly to the floor. Quickly grab both front legs with both hands and tilt the sheep into a sitting position on its tail at an angle of about 60° from upright. With practice the correct angle is soon found because if the sheep's body is too far forward it will surge back onto its feet, and if it is too far back it will struggle and kick. At the correct angle it sits quite relaxed. Modern mechanical restraining devices use this angle of tilt.

The above technique is especially useful for very large sheep or rams that are too heavy for the 'shearer's catch' as part of the Bowen shearing method[9]. Here the sheep is again approached from behind, grabbed by one hand around the neck and its front end quickly lifted up while walking backwards. The sheep can be kept walking backwards and a little more front-end lift will overbalance it and it will sit down on its rump at any desired spot.

To lift sheep up high off the ground such as over fences or into trailers, the main point is to use the Bowen method[9] of catching the sheep from behind so that the sheep's centre of gravity is kept high. The sheep can even be held with its front end raised, allowing it to stand on its hind legs. The thigh is then used to lever or knee the sheep onto the top of the fence while leaning back. Then it is simply rolled over the fence onto the ground, landing on its feet. This lifting procedure is illustrated in fig. 19.

The strain of lifting is best taken when the man's back is straight rather than bent.

Leader sheep

Sheep which are trained to lead through difficult areas where sheep are naturally reluctant to go could be exploited more by farmers. These sheep (Judas sheep) have been used in slaughter houses but there are many places on farms where they could be used, especially to assist loading sheep into trucks and woolsheds. A training pro-

High centre of gravity

1. Pull sheep up on its hind legs

2. Left hand holds right front leg

3. Right hand grips flank

4. Left knee pushes sheep over fence

Figure 19 Recommended way to lift a sheep to avoid back strain (source: Accident Compensation Corporation, Wellington, N.Z., poster)

gramme[10] for mature sheep has been developed. The main points are these:

- Pen the sheep regularly to get them used to a handler.
- Use a 1.5 m stick or pipe placed on the sheep's shoulder. When the sheep is used to this, slide the hand along the stick to touch the shoulder.
- In a small race, scratch or knead the shoulder area. While being scratched the sheep lifts its head high and flicks the tongue in and out of the mouth.
- Once the sheep is thoroughly tame and comes up to the handler for scratching, feed the sheep pellets while scratching it.
- Get it used to being tethered.
- Start the training programme by letting it follow you and the pellet bucket over the route it has to learn.
- Using the pellets as reward, teach it to open lift-up gate catches with its nose.
- Introduce simple verbal commands as the routine continues.
- Use the pellets as an irregular reward to retain its interest.
- Clearly mark the sheep to prevent it being sold or slaughtered.

Indoor sheep problems

In many systems around the world sheep are kept indoors for part or the whole of their lives[22]. It appears to take about 7–10 days for sheep to settle down after moving from outdoors to indoors. If the animals are to be put in individual crates (e.g. for experimental work) this will take longer and some individuals may never adjust and should be released.

Sheep penned as individuals or in small groups often engage in a range of idling activities that are sometimes described as depraved behaviour indicating stress. Examples are chewing the wooden rails, bar chewing, head banging, eating their own wool and the wool of their neighbours. Usually they are part of the adjustment process and in time their incidence declines. If they persist, action may be necessary to improve the environment or release the sheep.

Table 2 lists some of the recommendations from research about space requirements for sheep in intensive units in Australia.

Welfare

There are a number of key areas to consider in sheep welfare based on what is currently known about the animal's behaviour. These are discussed below in the form of basic recommendations.

Lambing

- The birth site must be respected. Do not move newly lambed ewes until bonding is assured.
- Orphan lambs should be mothered-on at the birth site if at all possible.
- Sheep need protection from predators.
- Avoid heavy concentrations of ewes lambing together. Stagger lambing to avoid mis-mothering.
- When recording is necessary at birth, use a means of temporary identification in the early morning to avoid disturbance. Tag when dam–lamb bonds have been developed a few hours later.
- Dogs used for lambing must be under complete control.

Handling

- Sheep need companionship. Handle in groups.
- They need to be able to see other sheep at all times.
- Keep the use of dogs to the minimum, and keep dogs under control.
- Handling facilities should be well-designed to allow good flow of animals.

Table 2 Recommended space needs of sheep in intensive housing (after Kilgour[17])

Facility	Space requirements		References
	adult	lamb	
Feed-lot			
hard surface	1.4 m²	0.56 m²	1
soil surface	2.8 m²	1.1 m²	1
	4.2 m²		
Grating floor pens			
(area per sheep)	1.0 m² or more		12
	1.05 m²		22
	1.1 m²		11
	1.1 m²		21
	1.4 m²	0.56 m²	1
Lambing pen			
small ewe	1.5 m²		1
large ewe	2.3 m²		1
Feed trough/rack			
(length per animal)			
individual feeder	10 cm		21
	15 cm	10 cm	1
group feeder	38 cm		21
	41 cm	30 cm	1
	50 cm	30–35 cm	12
	61 cm		22
height at throat	30–38 cm	25–30 cm	1
Water	0.01 m²/40 sheep		1
Holding pens	0.65 m²/ewe		1
Catching pens	0.3 m²		

- Sheep should be run through the facilities on occasions without any stressful or painful treatment.
- Yards should be checked for areas where sheep are hurt, bruised or frightened. Inspect the facilities from the sheep's viewpoint.
- Train labour to work without injury to themselves or the animals.
- The economic pressure to keep more sheep/man should not mean that lower standards of husbandry or welfare can be accepted.

Housing

- Correct space is critical. Although sheep flock, they still need their own individual space.
- Feeding must meet the animal's needs.
- Feed changes should be made slowly.
- Management should watch out for signs of abnormal behaviour. If problems arise because of boredom, their management and environment should be changed.

Drought

- Feed large amounts of food at infrequent intervals to allow all the animals to eat and reduce the competition.
- Teach young animals to eat drought rations before weaning so they will be pre-conditioned for any problems later in their lives.

Transport

- Animals must be pre-conditioned before lengthy travel on to their transit diet.
- Ventilation is critical.
- Adequate supervision of stock is essential at all times.

Shearing

- Newly shorn sheep are especially prone to cold stress. Adequate feed and shelter should be available after shearing.

Mutilation

- Farmers accept the stress of docking to prevent the greater stress of blow-fly maggots attacking dung-soiled sheep.
- Mulesing is carried out on wrinkled Merino sheep. The long-term cure is to select against wrinkles which were introduced as a fashion in the unfounded belief that they led to greater wool production.

References

1. Adams, J. (1973) 'Interest in sheep housing grows. The requirements of semi-confinement systems.' *Shepherd* 18(1), 14–15, 23.
2. Alexander, G. (1977) 'Role of auditory and visual cues in mutual recognition between ewes and lambs in merino sheep.' *Appl. Anim. Ethol.* 3, 65–81.
3. Alexander, G. and Stevens, D. (1981) 'Recognition of washed lambs by merino ewes.' *Appl. Anim. Ethol.* 7, 77–86.

4. Alexander, G. and Stevens, D. (1982) 'Failure to mask lamb odour with odoriferous substances.' *Appl. Anim. Ethol.* **8**, 253–60.

5. Alexander, G., Stevens, D., Kilgour, R., de Langen, H., Mottershead, B.E. and Lynch, J.J. (1983) 'Separation of ewes from twin lambs: incidence in several sheep breeds.' *Appl. Anim. Ethol.* (in press).

6. Alexander, G. and Williams, D. (1966) 'Teat-seeking activity in lambs during the first hours of life.' *Anim. Behav.* **14**, 166–76.

7. Allison, A.J. (1975) 'Flock mating in sheep: I Effect of number of ewes joined per ram on mating behaviour and fertility.' *N.Z. J. agric. Res.* **18**, 1–8.

8. Bourke, M.E. (1967) 'The mating behaviour of sheep.' *Wool Technol. Sheep Breed.* **14**(1), 77–80.

9. Bowen, G. (1978) *Wool Away: Art and Technique of Shearing*, Van Nostrand Reinhold Co., Whitcombe & Tombs, pp. 1–192.

10. Bremner, K.J., Braggins, J.B. and Kilgour, R. (1980) 'Training sheep as "leaders" in abattoirs and farm sheep yards.' *Proc. N.Z. Soc. Anim. Prod.* **40**, 111–16.

11. Bush, L.F. and Lind, R.A. (1973) 'Performance of ewes and lambs in confinement.' *J. Anim. Sci.* **36**, 407–10.

12. Cunningham, J.M.M. and Soutar, D.S. (1968) 'Sheep housing.' *Edinburgh School Agric. Misc. Publ.* No. 409, 6 pp.

13. Fowler, D.G. (1975) 'Mating activity and its relationship to reproductive performance in merino sheep.' *Appl. Anim. Ethol.* **1**, 357–68.

14. Hutson, G.D. (1980) 'An evaluation of some traditional and contemporary views on sheep behaviour.' In *Wool Harvesting Research and Development.* Proc. 1st Conf. Melbourne, 1979, pp. 377–85.

15. Keogh, R.G. and Lynch, J.J. (1982) 'Early feeding experience and subsequent acceptance of feed by sheep.' *Proc. N.Z. Soc. Anim. Prod.* **42**, 73–5.

16. Kilgour, R. (1974) 'Potential value of animal behaviour studies in animal production.' *Proc. Aust. Soc. Anim. Prod.* **10**, 286–98.

17. Kilgour, R. (1976) 'Sheep behaviour: its importance in farming systems, handling, transport and pre-slaughter treatment.' *Proc. Sheep Assembly and Transport Workshop.* West Aust. Dept Agric., Perth, pp. 64–84.

18. Kilgour, R. (1979) 'Hazards on hills for lambs at birth.' *Proc. 30th Ruakura Fms Conf.* p. 177.

19. Kilgour, R., Alexander G. and Stevens, D. (1982) 'A field study of factors interrupting bonding of ewes onto twins.' *Proc. 2nd Asia-Oceanic Congr. Perinatology*, Auckland (in press).

20. Lynch, J.J., Mottershead, B.E. and Alexander, G. (1980) 'Sheltering behaviour and lamb mortality amongst shorn merino ewes lambing in paddocks with a restricted area of shelter or no shelter.' *Appl. Anim. Eth* **6**, 163–74.

21. M.A.F.F. (1972) 'In-wintering the lowland flock.' Short term leaflet No. **63**, 1–8.

22. Williams, S. (1967) 'Indoor sheep – their behaviour, comfort and management.' *J. Royal Agric. Soc. Eng.* **128**, 46–52.

23. Zenchak, J.J., Anderson, G.C. and Schein, M.W. (1981) 'Sexual partner preference of adult rams (*Ovis aries*) as affected by social experience during rearing.' *Appl. Anim. Ethol.* **7**, 157–67.

4 Goat

The goat is one of the most widespread species of domesticated animal. It is found on all the continents and in the far corners of the world. It is one of man's most important animals and can be seen in the farming systems of the cold mountains, savannah brush lands, Mediterranean climates and deserts through to the tropics. It is also however, a threat to some ecological systems where its grazing habits have caused changes in habitat, erosion and the creation of deserts.

Behavioural definition

There is considerable confusion in the published texts about goats and sheep. Indeed sometimes they are considered to have similar behaviour and be grouped together. Because of this, the behavioural definition for goats is presented to contrast them with sheep.

Sheep and goats have different numbers of chromosomes and are clearly separate species. Crosses between them (described as geeps or shoats) are often reported but most are non-viable:

- The goat is more of a browsing animal than a sheep. They nibble and browse the leaves and bark of trees by standing on their hind legs. Their browsing habits enable them to cope with droughts which restrict grass growth. Long-legged breeds may kneel to graze.
- Goats are covered in hair (the Angora is an exception) and can browse scrub without becoming entangled. They can stand heat and humidity better than sheep.
- Goats are a lying-out species, where in contrast to sheep, kids spend long periods lying in undergrowth, feed infrequently and do not 'heel' and follow as lambs do. In adulthood they form smaller flocks than sheep, usually built-up from the extended family group.
- Goats move to shelter in wet and cold conditions more readily than sheep do. They have less surface cover and little subcutaneous fat compared to sheep. The short shiny coat of tropical breeds reflects sun and protects the goat against heat[11].

- Goats are more agile than sheep and are harder to restrain by conventional fencing. Goats jump more than sheep and dig holes below fences, a trait not common in sheep.
- There is a clear dominance structure in goat societies. In fighting, goats rear up and head clash with a downward stroke of the head and body. In contrast, rams move back and charge their opponents head on.
- Goats turn to face an attacker whereas sheep flee.
- Male goats display by masturbating and urinating along the belly and brisket causing a very pungent smell to attract the female. Rams do not do this although they may develop a strong smell at mating.
- Goats appear to develop a stronger bond with their owner or milker and are described as being more friendly than sheep. They are considered to be curious animals and rather capricious, smelling, examining, nibbling at everything that comes to their notice.

There is a lack of knowledge of the species-specific behaviour on the goat.

Sensory capacity

Goats have been shown to be able to distinguish grey from three bands of colour. These are light red-violet through red to yellow, yellowish green to green, and blues. They appear to respond best to orange and worst to blues. However, goats show a much greater reliance on smell and odour than vision in their environment.

The goats' sensitivity to various sounds is apparent from the range of snorts, foot stamping and bleats which they use in communication. The goat alarm call is demonstrated best by the buck's spitting-like call, jerking action of the head and foot stamping. Studies[20] in Hawaii showed that when this call was made all other goats scrambled to their feet and repeated the behaviour. The males vocalise frequently during courtship as do does and kids during active grazing together.

The alarm call when given by the does causes kids of up to a week old to drop in the undergrowth while a muted vibrato sound will call the kid from its lying-out position. Alarm calls of kids are high pitched wails that often sound like a child screaming.

Birth

Behaviour around birth

Just before birth a doe is often described as more fretful and nervous. Hollows can be seen in the goat's flank and pronounced hollows on either side of the tail. Sometimes there is a slight mucous discharge from the vulva but this may change over a day or so to be opaque or slightly yellow. If indoors, the doe will paw the bedding and as birth draws near she will repeatedly lie down and stand up with signs of straining. About 24 hours before birth, the udder of milking goats will swell markedly. However this time interval can vary considerably.

Few births have been witnessed and documented from wild or feral goats but indoors, birth is usually complete within about 1 hour. Litter sizes may be from one to four kids but in feral goats that suckle their young, there is usually a high mortality in multiple births and singles are usually reared. However, performance is very dependent upon the environmental conditions. Afterbirths are passed from about half an hour to 4 hours after birth is finished and they are often eaten.

The initial response of the kids is to seek a rounded ventral surface without hair. When a hairless area is reached they start sucking movements. When an additional udder was transposed to the neck of a doe kids quickly found it and used it[17].

Kids quickly establish themselves on a preferred teat and there is very little non-nutritive sucking, i.e. sucking in the first few days of life without a need for milk. During the strong suckling periods it was observed that the movements of the kid's tail corresponded exactly with the frequency of sucks during the major peak[19].

Dam—offspring bonding

To achieve a good bond between doe and kid it is important that no interruptions disturb them immediately after birth. Studies[3] have shown that a metronome ticking in the long grass near a doe during birth was enough to disturb the bonding and more kids from disturbed does died before weaning than undisturbed controls.

In poorly bonded animals the doe finds it difficult to locate the kids, and in turn the kids are not calmed by the presence of their mothers. This has been shown in kids at 2 months of age that were born in disturbed and non-disturbed birth situations. The calming of a young animal after rejoining its dam is a practical sign of good bonding. Figure 20 shows the typical response of does accepting their own or rejecting alien kids.

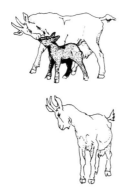

After
5 mins
contact
at birth

Kids accepted after
1–3 hr separation

Distress on removal

Without
contact

Doe aggressive or indifferent

Figure 20 Reactions of does to kids after limited contact at birth (after Klopfer, P.H., McGeorge, L. and Barnett, R.J. (1973). *Addison-Wesley Module in Biology* No. 4, pp. 21–4)

Feral does with kids keep together by bleats especially when they are actively moving and grazing. Kids have been observed to bleat once every eight seconds but during rest silence is maintained at all times.

Investigations[7] of bonding of does and kids showed the importance of means of early recognition of the doe's own kid while alien kids are rejected. This has been called 'maternal labelling' where the initial licking of the kid after birth not only dried it and stimulated breathing but also appeared to label the kid so that the dam could recognise it later. Experimental tests clearly showed that the licking and suckling of a kid by the doe was vital for later recognition and a good bond.

A study[14] of feral goats described their lying-out behaviour and showed that the does left the kids for 2–8 hours, and later returned to the precise position where they left them. A muted vibrato call was given and the kid ran to her to nurse. She nuzzled the perianal area of the kid on meeting. At night they kept very close together.

During the lying-out period which one study[14] said lasted 3 days but others say lasts up to several weeks in open country, the kid rises and stretches every few hours and then lies down again. During the very early lying-out period after birth, human disturbance does not alarm the kid but later on they are easily frightened and run off. If the doe gives the alarm call to kids less than 7 days old, some of them sink to the ground and freeze. If the kid gives the alarm, the doe quickly moves to defend it against intruders including males and castrated male goats.

This behaviour highlights the value of a good dam—offspring bond for high survival. It also shows the importance of the dam knowing its grazing territory. Thus in farming goats it is advisable not to move them to a new area just before kidding, and that the kidding area would be better if it had some landmarks (hills and hollows) rather than be flat and uniform. It should also improve survival if there was ample cover safe from predators for the early lying-out phase after birth.

Present-day recommendations for herding based on experience in large herds in the United States of America are that the does and kids should either be given intensive care, or that the herd should be left alone entirely (easy-care). Anything in between does not seem to be successful in terms of good weaning percentages (kids weaned per 100 does joined with the buck). This approach agrees very much with experience from sheep systems in Australasia.

Rearing

Kids reared naturally on their dams appear to have few behaviour problems. As they feed infrequently compared to lambs, artificial rearing systems are usually easy to organise.

Kids can be bottle fed small amounts of milk (400–600 ml) about 5 times per day after birth declining over a few weeks to three times or twice a day with weaning from milk at 2 months of age. Bottle feeding is very similar to recommendations for lambs and as in all species, having the colostrum from its dam is essential to receive immunity to local diseases. If this is not possible, colostrum from another doe, ewe or a cow can be used. As cow colostrum can be obtained in larger quantities, it can be frozen in separate small containers for when needed. Cow colostrum is a last resort.

It has generally been recommended that milk should be fed to kids at blood heat but recent developments of rearing kids on cold milk on *ad lib.* feeding systems have been very successful, achieving good growth rates with freedom from digestive upsets.

Some farmers recommend that buck and doe kids be separated at 3 months old, to give a higher energy feed to improve growth and exercise to avoid excess fatness and lethargy in the males. Where large groups of males are run together they are usually castrated to increase growth rate and avoid the development of pungent odours.

In Mediterranean climates (Australasia) where there is a growing interest in goat farming, to achieve high kid survival mobile shelters are provided at pasture during the rearing stage[8]. Studies[10] of feral goats showed that after kidding in the tree areas, the doe brought the young out after about 4 days and they remained some distance from the group of other does for up to 3 weeks. During this time the kids started to play together in a group but still maintained closeness to their dams. The doe would butt away alien kids and other does. The males tolerated the approach of the kids, even when they played around them, climbing over them while the males were resting and ruminating. The association between dam and kid lasted after weaning until just before the birth of her next kid when the older kid was driven away.

Goat kids start to nibble grass from 3 weeks old and after 8–9 weeks are effective grazers. The age of weaning in the wild depends on the dam's milk supply but on farms they are weaned at about the same age as lambs (12–14 weeks).

Cross fostering with sheep

Fostering kids on to ewes and lambs on to does is an old established practice. Kids have never been fostered on to cows. Some farmers claim that kids reared on ewes develop an improved attitude to eating grass from their early experience. However more information is needed about these practices before they can be recommended with confidence.

Some deliberate cross-fostering experiments[18] have been done where a doe reared a lamb along with her kid and a ewe reared a kid with her lamb. The ewe kept losing her kid as it would lie-out for long periods and she would bleat and seek for it. The lamb fostered on the doe, trailed her constantly and with frequently feeding soon outgrew her kid co-twin who spent a lot of time lying-out.

When the age was reached to settle the social interactions between the ram lambs and buck kids, the lambs backed off to charge while the kids rose on their feet to head-clash. The kids were thus butted in the belly and the lambs soon became dominant.

Reproduction

The male

Young males reach puberty from about 4 to 5 months old and will start sniffing the perianal area of any doe (young or old) as she approaches oestrus. They are usually prevented from actual mating by older bucks.

Males tend to congregate in separate male groups ranging from three to eight in Hawaii[20]. With males being harassed by hunters and predators larger groups form for short periods. These groups persist for a large part of the year but break up and the males move into the family groups of does and kids during mating. Figure 21 shows the pattern of male movement into and out of doe groups during the breeding season.

As the breeding season approaches, the mature males engage in scent urination of the hair. Here urine is directed onto the body rather than into wallows or areas of their territory as in other species. At the start of this clearly defined 'rut', the males stop urinating in the usual way and start squirting a thin stream of urine along the belly, chest, brisket and into the beard at high pressure from the erect penis.

This is achieved by either arching the back and bending the hind legs to allow urination on chest and front legs, or by arching the body and bending one hind leg to allow urination into the mouth, beard and throat. Up to 7 cm of the penis may be taken into the mouth. The penis pulses in a vertical plane during urination, and afterwards the buck gives the lip curl and yawns. Does have been observed[4] to approach bucks and show continued interest in them after they have performed the above behaviour. Another scent gland near the horns is also active during the rut.

Scent urination may occur during courtship or when the males are present. How increasing age or dominance affects the smell of bucks is not known.

Behaviour studies[15] have described the changes that occur with the onset of the rut. These are concerned with:

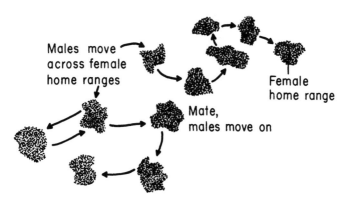

Figure 21 Pattern of male movement in female range areas during rut[16]

Hunt - sniff

Lip-curl (flehmen)

Figure 22 Pattern of responses leading to the lip-curl or flehmen[16]

- Spreading of odour. This includes scent urination as described together with elevation of the tail to assist the spread of scent from the tail gland.
- Physical display through posture. These can be either the head held high above the doe, or a head-down display when the male follows the female with muzzle close to the ground, neck outstretched and back arched. Shoulder and neck hairs are also raised. The same display occurs during dominance interactions in males.
- Testing the female's oestrus. The buck approaches and sniffs the doe's perianal area and she either moves on or urinates before moving on. She may also turn and head-threat the buck. The buck may sniff the urine on the ground then lift the head and give a lip curl after which she is either ignored or followed (see fig. 22).
- Courtship. Courtship response shown by the male before mating can be a low stretch where the head is held low and horizontal while being stretched forward. It can be a twist where the head is rotated as the buck moves towards the doe, and it can be a stiff-necked kick or paw with the front foot at the female seldom striking her. The buck may flick his tongue in and out while making a low grunting sound. These various courtship displays may be given singly or in any combination, but the lick rarely occurs after a low stretch. Older males give more intensive displays than younger males. Bucks also vocalise a lot during courtship.

Training males for AI

From experience feral bucks and males from five dairy goat breeds have been trained to serve into an artificial vagina using either oestrous does or bucks as teasers. Training is best begun with a restrained oestrous teaser doe, and later a restrained male will be sufficient stimulus for mounting. In the non-breeding season an oestrous doe may be required from time to time to re-stimulate males which refuse to work.

Feral males took from 2 days to several weeks to train and a few bucks could not be trained. In the breeding season dairy breed males all responded within three exposures even to restrained males.

Once trained, males can be collected once or twice daily in season and some continue to be active throughout the year in temperate climates.

The female

Females reach puberty between 4 and 6 months old and in the wild or feral state they will be mated at this time. In some farming situations, mating may be delayed until 10–20 months of age. U.S.A. and European commercial dairy goat farmers often mate the does at 12–14 months, when body weights have reached about 66% of that of the mature female. In temperate climates females cycle every 19–21 days and stay in oestrus from 1 to 3 days. However, there can be periods of more intense oestrous activity during the year, usually in the autumn in temperate climates, but there may be another peak of activity after kidding in spring. Females are mainly in an oestrum between spring and autumn. In the tropics local breeds of goat show oestrus throughout the year while introduced goats like the Angora show a persistent three-months breeding season.

Most range observations have shown how goats mass into family groups in a limited home range area of 0.75–1 km². The female group is usually made up of a doe with her kids and juveniles, a total of three to five animals. If groups have been separated on farms for as long as 6 months, when recombined they move back into their family groups.

Feral does shot as noxious pests are regularly found to be pregnant while lactating, so lactation is not a hindrance to oestrus which it is in most sheep breeds and the suckled cow. The goat is renowned for lactating for a very long time without requiring further pregnancy. In some heavy-milking dairy strains, they will lactate without having been mated at all.

Does in oestrus vocalise a lot at night and generally take an interest in any males that are present. Tail-wagging is a key sign of oestrus. There is very little female–female mounting.

Social dominance and mating

The most dominant male in a group (called the alpha male) carries out most of the mating. In feral studies[10] the oldest male in the herd gave twelve mounts to three does while the next oldest was only able to mate once. Each receptive female was dominated by the alpha male, who also head-threatened or made a few rapid strides towards other males to exclude them. Social dominance in a group of five males corresponded well to body weight and horn length. The alpha male tolerated the young, 6-month-old males sniffing the does, and although many males chased the doe coming into heat, when she was fully receptive only the alpha buck mated her.

No studies have been reported which give a buck:doe mating ratio for successful range or pasture management systems. Farmers tend to assume the same mating ratio for sheep, i.e. one buck to forty to fifty does.

Social dominance is clearly altered by de-horning which should be done before 2 weeks of age if possible. However, it may be necessary to modify social dominance in mature (feral) animals by de-horning. Veterinary advice should be obtained. The shock of de-horning mature pregnant does has been known to cause abortion, and adult de-horning exposes large frontal sinuses that take time to heal.

Production

Most of the world's 400 million goats are found in the tropics and are usually kept in herds of two to eight animals by small holders and villages. More goat meat (chevron) is eaten in the world than mutton and goats also provide milk, cheese, hair and skins. An extensive review of the goat's role in world farming can be found in reference 6.

Free-range systems

The main problem with farming goats on free range systems is usually that of fencing to restrain the animals. Goats are agile and can climb up the stays that support fence posts and jump over. They can negotiate eight-wire fences that normally hold sheep and they can find and negotiate holes in and under the fence. They will even dig at and expand existing holes. Barbed wire may slow their escape but inflicts severe damage to hides and udders. The bottom tight wire must be within 100 mm of the ground.

Electric fencing or electric wires on the top and bottom of standard fencing are successful but pre-conditioning of the goats is important so they recognise and respect electric fences. The main recommendations for electric fences for goats are shown in in fig. 23.

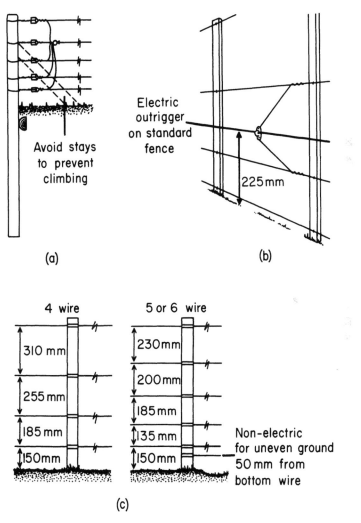

Figure 23 Details of suitable fences for goats. (a) All-electric with no stays; (b) electrified out-rigger on fence; (c) four- or five-wired fences (source: Gallagher Electronics, Hamilton, N.Z.)

Four days inside a confined yard were found[13] to be needed to train goats to respect electric wires. Electric fencing units that give irregular pulses are very useful for goats.

Conventional fences in New Zealand have been found inadequate to farm feral goats especially if they are being used for weed control. It has been suggested by farmers that goats should be shifted regularly to prevent them setting up their own movement patterns when they will use their ingenuity and skill to escape. Once goats have come to 'know' their home territory they cease to wander when farmed for meat, hair or milk.

Woven mesh wire of 80 cm squares is goatproof but very expensive. In tropical areas with small herds a small bamboo yoke is often put round the neck of a goat to prevent it from escaping or gaining entry on farms. A yoke of heavy wire, plastic pipe or timber can be used.

The ingenuity of goats can be used to trap them. This is done by making a walkway over the fence (1.4 m high) to allow the goats to jump in to the property but which is too high for them to jump out of (fig. 24). A small artificial precipice-type drop can also be built in the fence line. Similar practices have also been developed to trap deer.

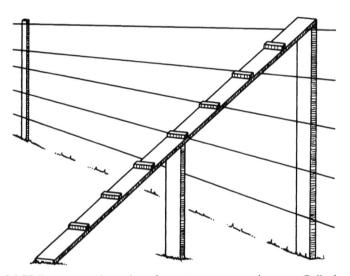

Figure 24 Walkway over boundary fence to trap goats (source: Gallagher Electronics, Hamilton, N.Z.)

Normally in the tropics goats are coralled at night to prevent predation and theft. These advantages have to be balanced against the loss of production from the reduced feed intake the practice causes.

Indoor stalling
Many goats, for example in Java, are kept in small bamboo hutches and grass and 'browse' is brought to them daily. The housing is either

on the ground or on stilts and along with bedding that has to be provided, the costs of the enterprise can be relatively high. Zero-grazing or stall-feeding is very common in well-managed dairy goat herds in Europe (1.5 m² /animal is recommended).

The goats' behaviour may be affected by the husbandry system where lethargy and obesity may result from lack of exercise or un-balanced diet, and feet problems may arise through excessively long hooves.

Tethering

Tethering of goats is commonly practised where control of the animal is needed to prevent it from wandering and damaging neighbouring crops. By tethering, a goat can be forced to browse weeds or other undesirable plants. Care must be taken to prevent tethered goats becoming tangled with each other.

The provision of water may be needed but shelter is more essential. Regular changes of the tether which contains two swivels and a broad comfortable collar are also essential. A running tether − chain on a wire between two stakes − is preferable to a fixed tether.

Early handling of kids and having them imprinted onto humans as well as goats makes regular handling and tethering easier.

Feeding

A growing number of studies report the grazing behaviour of goats. On Texas range[1] 28% of the 24 hour day was spent grazing grasses and weeds. This was similar to feral goats in N.Z. where 30% of the day feeding was recorded[10]. The daily grazing rhythm was also similar in both studies with a grazing period from dawn to 10.30 a.m., then rest followed by grazing from 11.00 a.m. to noon. This was followed by a rest then grazing from 14.30 to 15.30 after which the animals rested again. There was another main grazing period for 3 hours before sunset. In all, 30% of the day was used in feeding, and of the total feeding period 34% was grazing and 65% browsing. Ruminating took up 10% of the total time with 46% resting. Most resting took place in winter and least in autumn. About 12% of the goats' time was spent travelling about 3 km/day while grazing or going to water. The animals were seen to defecate on average 3.4 times and urinate five times per day, but these figures will be low.

In studies[2] with penned goats, most of the rumination (75%) occurred at night with only brief periods of rumination during the day. However, there was very wide variation in ruminating times between individual goats in the study.

Goats eat a lot of roughage which includes weeds, woody shrubs, leaves and bark. If given free choice with sheep they compete over herbaceous plants but eat less grass and legumes. Thus in savannah or scrub country the two species are not in direct competition. Goats will however thrive on pasture species with a supplement of hay provided. In the main grass-growing season hay may not be eaten. Areas of their range need to be checked for poisonous species which they will eat. Many flavours in goats' milk can be caused by the plant species they are eating. They will also nibble gates and fences so lead paint should be avoided.

Goats have a habit of eating a plant from tip to base, eating off seed heads and flowers and leaving the regenerative parts of the plant. Their mobility also causes seeds of browsed species to be widely spread. Where chemical weed control by herbicides is too expensive, intensively stocked goats are often used.

A number of tropical studies[5] have shown the superiority of goats over sheep in the efficiency with which they digest cellulose and lignin.

Shelter

Studies of the sensitivity of goats to heat and cold showed that mature pregnant females could withstand cold for longer periods than young animals. In a cold chamber where goats were kept at 0.5°C, with wettings every 30 minutes, 40% of adults withstood the cold for 48 hours before collapse whereas the group of five young goats collapsed after 12 hours. There were no abortions in the pregnant goats.

In normal tropical conditions, studies[9] showed goats to have a higher rectal temperature and heart rate than either cattle or water buffalo.

Thus shelter to provide warmth, shade and protection from rain is important for goats. Mohair goats after shearing require special care. About 0.5 m^2/animal is adequate sheltered area for goats.

Milking

Milking machines are now accepted as part of modern goat dairying and the goat soon adapts to being machine milked in a range of facilities such as herringbone and rotary parlours. In some systems concentrates may be fed at milking time in the parlour. Where this occurs, baffles are recommended to prevent goats reaching to steal their neighbour's feed, nibbling their neighbour and horn butting thus upsetting the dairy routine. It is important to have goats trained to the milking shed and handling routine before the milking season begins. This is especially the case for first-lactation animals.

Intensively farmed dairy goats are subject to problems such as mastitis, vaginal prolapse and pseudopregnancy. Although lactation will continue for a long period, a dry period of 8 weeks between lactations is recommended.

Fibre production

The main behaviour problems arise with shearing. Goats, especially kids, can cry loudly when being held for shearing. The shearing handpiece is often set at 1500–2000 r.p.m. for goats. This is slower than for sheep shearing (2400 r.p.m.). As the hair has no grease in it, the handpiece may overheat and cause some stress to the goat. A dip in one part of lubricating oil and two parts kerosene will cool combs and cutters[12].

In places the hair lies close to the skin, especially in the neck and shoulder areas. These areas should be shorn against the lie of the hair, otherwise goats are shorn in the same manner as sheep.

General handling

Goats can generally be handled in range conditions with dogs as are sheep. Indeed having sheep in with feral goats may improve their flocking and mob control. Before mustering it is useful to know their established movement pattern over the range and muster to break that and confuse them. Once on new, unfamiliar territory they are more confused and are easier to handle.

Too much pressure on a mob will cause old bucks and does with kids to fight the dogs and break away. Animals that have broken away are very difficult to recapture on their own and it is best to allow them to become part of a group again. There is a risk then of losing the whole group, led by the escapee.

Handling goats in pens or yards is very similar to sheep except that they jump over the fences designed for sheep. Great care must be taken the first time goats are worked, and it is often preferable to handle them in facilities built for cattle or deer with very high fences. The same principle for flow through handling facilities outlined for sheep apply to goats. The large horns on goats often make them nervous of narrow, drafting races from fear of being caught.

Welfare

From what is known of the behaviour of the goat, the following are some recommendations aimed at the animals' welfare:

- Goats need shelter from cold, heat and heavy rain.
- Good fencing is required to restrain goats. The fence must restrict their movements without causing injury.
- De-horning of farmed goats should be done with veterinary advice. It will help prevent injuries to other goats and their handlers. It will alter the social status in the herd however. Disbudding at 4–10 days old is recommended.
- Housed goats must be given dry bedding and milking goats require care and good hygiene to avoid udder problems.
- Good animal health programmes are essential along with adequate nutrition to meet the animals' production needs, especially those of lactation.
- Some roughage or browse is beneficial to goats.
- Goats should not be disturbed while kidding.
- The flocking of range goats can be improved for handling purposes if some sheep are mixed with them.
- Goats are not as dependent as sheep on having other goats around them so they can be worked reasonably well singly or in small groups provided they cannot escape by jumping.

It is very easy for people closely associated with small groups of goats to become very anthropomorphic about them and their needs. As with all animals the variation in behaviour between individuals is large and conclusions from individuals should not be blindly applied to groups, especially those run in larger commercial operations.

References

1. Askins, G.D. and Turner, E.E. (1972) 'A behavioural study of angora goats on west Texas range.' *J. Range Mgmt* **25**, 82–7.
2. Bell, F.R. and Lawn, A.M. (1957) 'The pattern of rumination behaviour in housed goats.' *Anim. Behav.* **5**, 85–9.
3. Blauvelt, H. (1956) 'Neonate–mother relationship in goat and man'. In "Group Processes". *Trans. 2nd Conf.* (ed) B. Schaffner, Josiah Macy Foundation, pp. 94–140.
4. Coblentz, B.E. (1976) 'Functions of scent-urination in ungulates with special reference to feral goats (*Capra hircus L.*).' *Amer. Nat.* **110**(974), 549–57.
5. Devendra, C. (1978) 'The digestive efficiency of goats.' *Wld Rev. Anim. Prod.* **14**(1), 9–22.
6. Gall, C. (1981) (ed.) *Goat Production*. London: Academic Press, 1–619.
7. Gubernick, D.J. (1980) 'Maternal "imprinting" or maternal "labelling" in goats?' *Anim. Behav.* **28**(1), 124–9.
8. Holst, P.J., Carberry, P.M., Mitchell, T.D., Trimmer, B.I. and Kajons, A. (1976) 'Husbandry of bush goats.' *N.S.W. Dept Agric. Publ.* pp. 1–18.

9. Jindal, S.K. (1980) 'Effect of climate on goats: A review.' *Indian J. Dairy Sci.* **33**(3), 285–93.

10. Kilgour, R. and Ross, D.J. (1980) 'Feral goat behaviour – a management guide.' *N.Z. Jl Agric.* **141**(5), 15–20.

11. Macfarlane, W.V. (1982) 'Concepts in animal adaptations.' *Proc. 3rd Int Conf. Goat Prod. Disease*, Tucson, U.S.A. pp. 375–85.

12. Mitchell, T. (1977) 'Angora goats.' *Agric. Gaz (NSW).* **88**(2), 5–8.

13. Niven, D.R. and Jordan, D.J. (1980) 'Electric fencing for feral goats.' *Q. Agric. J.* **106**(4), 331–2.

14. Rudge, M.R. (1970) 'Mother and kid behaviour in feral goats (*Capra hircus L.*).' *Z. Tierpsychol.* **27**, 687–92.

15. Schaller, G.B. and Laurie, A. (1971) 'Courtship behaviour of the wild goat.' *Z. Saugetierkunde* **39**, 115–27.

16. Shank, C.C. (1972) 'Some aspects of social behaviour in a population of feral goats (*Capra hircus L.*).' *Z. Tierpsychol.* **30**, 488–528.

17. Stephens, D.B. and Linzell, J.L. (1974) 'The development of sucking behaviour in the newborn goat.' *Anim. Behav.* **22**, 628–33.

18. Tomlinson, K.A. and Price, E.O. (1980) 'The establishment and reversibility of species affinities in domestic sheep and goats.' *Anim. Behav.* **28**, 325–30.

19. Wolff, P.H. (1973) 'Natural history of sucking patterns in infant goats; a comparative study.' *J. Comp. Physiol. Psychol.* **84**(2), 252–7.

20. Yocom, C.F. (1967) 'Ecology of feral goats in Haleakala national park, Maui, Hawaii.' *Amer. Midl. Nat.* **77**(2), 418–51.

5 Deer

Deer are now being farmed in several countries of the world, where previously they have run wild. In New Zealand they were once classed as a noxious pest. Farming systems have been developed' in which a knowledge of deer behaviour has played an important part. This chapter summarises the available information. Although many deer strains or breeds are being farmed, this chapter is based mainly on information about Red deer (*Cervus elaphus*) and Fallow deer (*Dama dama*).

Behavioural definition

A deer is a medium-large, hair-covered, browsing ruminant that shows strong seasonal patterns attuned to body growth, antler growth and mating (rut). It lives on the verge between open forest and grassland where it finds food and shelter. It shows a strong attachment to a home range and depends upon acute hearing, smell and speed to avoid predators and capture. Outside the rut the sexes are organised into their own social groups but nearing the rut the groups become more mobile and social stability breaks down. The young are born rapidly, they stand and suckle soon after birth before lying out during the first few weeks of life during which time they are fed infrequently by their dams. After this time, they join their dam's group and slowly disperse between 1 and 4 years of age.

Terminology

Terminology varies with class of stock and breed. For example:

Juveniles	Calf:	Red deer or Wapiti
	Fawn:	Fallow deer
Mature females	Doe:	breeds other than Red deer
	Hind:	Red deer
	Cow:	Wapiti
Mature males	Buck:	Fallow deer (they roar or grunt)
	Bull:	Wapiti (they bugle)
	Stag:	Red deer (they roar)

In this chapter, the terms calf, hind, stag and roaring are used for both species (Red and Fallow).

Birth

Hind behaviour before birth

In late pregnancy hinds spend more time resting. A few days before calving they become restless and farmed deer pace the fences. Some become aggressive and bellow more frequently. The hind's swelling udder and vulva are signs of approaching birth and they will separate out to the fringes of the herd and into a quiet place if this is available, from 2 to 22 hours before giving birth. They prefer higher ground for calving.

Hind behaviour during birth

With the onset of strong contractions, the hind stops grazing and settles in one area alternatively standing or lying. With heavy contractions she rolls and strains heavily. After the calf is born, some hinds rest for up to 5 minutes and then on rising lick the calf and all excess birth fluids. The afterbirth comes away from 2 to 5 hours after the calf is born and this is carefully eaten by the hind.

In farmed deer in New Zealand, average birth weights of calves are Reds 7.5–9.5 kg, Fallow 3.6–4.5 kg. Few difficult births have been observed but mortality is highest among the very large calves (above 10 kg) and the very small. Mortality may increase with crossing larger sire breeds (e.g. Wapiti) on smaller breeds. Hinds giving birth for the first time (first-parity dams) show higher calf mortality.

Hind and calf behaviour after birth

After being licked by the hind the calf stands up about 30 minutes after birth and starts to move and squeak. The time between birth and starting to suckle varies from 10 to 129 minutes.

After suckling, the calf moves away from the hind and drops down in the grass or bushes. This is a behavioural characteristic of the lying-out species which is especially developed in deer. The calf appears attached more to its birth site in the early stages of life although some move around with their dams. This behaviour is later expressed in wild deer by their ability to home to a 'pad'. Calves are well adapted through camouflage to lie in hides unseen by predators.

In Red and Fallow deer after the calf has taken up its lying position after birth, the hind usually grazes within 50 m. As the calf grows older, the hind will then graze further away from the hide, even as

far as 1 km away. Observations[13] showed that hinds visited their
calves to suckle them on average 2.8 times/day during the first week
and this dropped to once a day during the fourth week after birth.
The length of these suckling sessions averaged 149 seconds in the
first week and 50–80 seconds in the second week. While suckling the
calf moves from one teat to the other (deer have four teats) and
observations[1] showed that some sucklings could last up to 4 minutes.
Milk intake of fawns was recorded as 150–630 g/day in early lactation which peaked at 1400–2000 g/day soon after birth. Calves
started to nibble grass by 3 weeks old and were grazing regularly at
6 weeks of age. Some well-fed hinds have produced 140–180 kg of
milk in a 150-day lactation.

Hinds are very alert and nervous as they approach their hidden
calves for suckling, compared to their behaviour during grazing.
During suckling the calf is stimulated to urinate and defecate by the
hind licking its perianal area. The hind then eats these waste products
and this behaviour continues for the first 2 weeks after birth. However, the hind will keep on licking the calf's anogenital area for up to
6 weeks after birth.

While the dams are in contact with their calves for suckling, the
pair may go for very short excursions usually within 3 m but
occasionally up to 30 m. When calves leave their dams to lie down
again, they usually walk away no more than 20 m. After about 2–3
weeks of age the calves follow their mothers and become settled into
normal herd life.

There is considerable social communication between hinds and
their calves. Hinds bark at other hinds warning them not to come too
close to their lying-out calves. The high-pitched, piping notes of the
calves are answered by a muted mewing-lowing sound of the hind.
Some Red deer mothers will charge in response to wailing made by
calves more than 2 days old and both Red and Fallow will bark
loudly when surprised. Fallow hinds may combine barking with hoof
pounding which spreads the alarm to others. Lifting the tail to
expose the underside is a silent alarm signal in Fallow.

Handling calves

In farming situations, newly born calves may have to be handled for
management reasons and this can disrupt normal behaviour patterns.
If calves are handled in the first 2 days after birth they normally
'freeze'. Studies[5] have shown that the normal resting heart rate is
140–170 beats/minute which reduces rapidly during freezing
behaviour to 50–60 beats/minute. This returns to normal 1 minute

after leaving them alone. After 2—5 days old, the calves run off when disturbed.

Other studies[13] showed that in tagging and weighing forty-five Red calves on the first day after birth, all but two stayed still on the approach of the stockmen. Half of them squealed on tagging and although this alerted their dams, they did not come closer than 30 m. About 70% of the calves then lay down after handling, the remainder running away. The calves used the shelter available (long grass, branch hides and scrub) after the first day. They ran away from the stockmen's approach on the fifth day from scrub, on the seventh day from long grass and on the ninth day from the hides made from branches. They joined the herd with their dams on the twelfth day.

On rare occasions after human handling, hinds have been observed to take the calves in their teeth and flail them with their forelegs. This behaviour may be seen in newly captured deer in a strange environment such as a densly stocked area with little cover for the calves.

If calves have to be handled for example for tagging this should be done on the first day after birth. Gloves may reduce the human smell left on calves but some stockmen find this makes little difference. The calves should be forced to lie down or freeze in the area they were found in, before the stockmen leave. Most farmers leave deer strictly alone during calving due to the risk of higher mortality through handling and subsequent mis-mothering. For simple recording, calves are tagged soon after birth and the dam—offspring pairs are noted using binoculars after they have mothered up.

Rearing

Hand rearing orphans

The first priority with any orphan is to give it colostrum as soon as possible after birth, otherwise chances of survival are low. This colostrum should be from its dam, another dam, a sheep, goat or a cow, in that order of priority. Orphans may be dehydrated and an injection of sterile glucose saline might be needed. Seek veterinary advice.

Milk substitutes should be as similar as possible to deer milk. Red deer milk is 22% lactose, 33% protein and 40% fat which is higher in fat and lower in lactose than the ewe or cow.

Calves drink from 5 cm long teats with a 1 cm slit rather than a hole. When group fed the young should be of similar size. They will approach and suck well from a suspended bucket with teats inserted.

Initially warm milk (30°C) should be offered but after training cold milk will be readily accepted.

For the first week Red calves should be fed four times/day and then three times/day for the following 7 weeks. Calves at 2–3 days old will drink a total of 400–800 ml/day increasing to 2 litre/day from 2 to 8 weeks. Fallow calves need about half this quantity daily. Calves are taught to suckle from a teat by pressing the milk from the teat with a finger, or sucking a finger dipped in milk. They can be taught to drink from buckets from a few days old.

Calves need a draught-free, dry environment, with bedding changed regularly. Some artificial warmth is helpful. Care must be taken to wipe the anogenital area to stimulate the calf to urinate and defecate and this must be continued for a number of days.

Some concentrate feed can be offered from the third week at which time good clean pasture should be available. (See references 6 and 17.)

Effect of human handling

Early human handling can modify the behaviour of adult deer. Research[9] has shown that hand-reared Fallow deer lie further apart at pasture than semi-wild deer. Wild hinds bite others in competition over food or will bite and kick strange calves. This behaviour was greatly intensified in hand-reared animals.

Bottle-fed calves have been shown[9] to be strongly attached to their human caretakers and when tested at 4 years of age did not mix with other deer and were aggressive towards them. However, general experience is that hand-reared calves soon integrate with wild deer and show normal reproductive behaviour. Most practical concern is with hand-reared stags that have lost their fear of humans and can be very dangerous during the rut. Hand-reared male calves are best slaughtered at 15 months old or castrated about 3–6 months of age before any antler development occurs and they become sexually active. Hand-reared calves can be trained to come to a call over distances up to 1 km. These animals are then used to help lead and move groups of animals in farming situations.

Removing the fawn from a hind does not seem to cause her very serious stress like that seen in a follower-type species such as sheep. Studies have shown that hinds keep up their normal eating patterns and stop calling for the calves from about 30 hours after removal.

Studies[9] have shown that if deer are reared from an early age by humans in isolation from other deer, they have an altered response to their own species and their behaviour is very different from that of tame farmed deer.

Weaning

In wild deer, calves appear to wean themselves at about 20 weeks of age but some will keep on suckling after 35 weeks. depending on the dam's milk supply. During this long period with the dam, the calves learn details of the environment, and in particular they follow their feeding patterns.

In farmed deer, weaning usually takes place around 8–12 weeks of age at 30–35 kg liveweight. These weaners should be introduced to the pastures where they are to live and should be carefully worked preferably without dogs, calling to them by voice. This area will then become their 'pad' or home range. An ideal area should have a variety of diet, shelter, shade, water and opportunity to wallow for Red deer and freedom from harassment. Newly weaned calves should be separated by a visual barrier from the hinds so they will not run into fences and injure themselves when trying to return to their dams. Deer farmers are currently using both set-stocked and rotational grazing systems.

Veterinary advice should be obtained to provide an animal health programme for weaners, especially when stocked intensively. Deer are prone to the parasitic diseases of cattle and sheep. Wallows which are not being used may indicate unhealthy Red deer, although animals in high fever after difficult births may spend all day in the wallow.

In some farming situations identifying offspring with their dams is left until weaning when the following routine is used:

- At 5 p.m. wean calves from hinds, tag them and hold them in quiet dark confinement overnight.
- At 8 a.m. let them out in small groups back to the hinds.
- Using binoculars record the dam and offspring tag numbers.
- The next day, wean them permanently.
- Weaning before the rut allows the dams to be selected for mating based on their performance (e.g. weight of calf weaned).

Reproduction

The male

The seasonal behaviour and social life of deer is one of dramatic

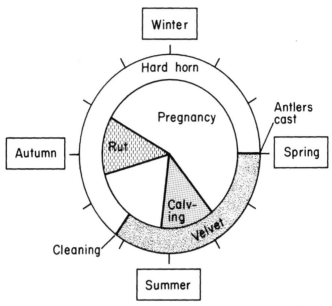

Figure 25 Seasonal changes in the life of red deer[16]

change and this hinges very much on the stage of the male's antlers as shown in fig. 25.

Antlers are not horns but are laid down as cartilage which is then converted into bone. Antlers grow from the pedicle which is part of the skull, and are controlled by the hormone testosterone.

Research[18] has shown that deer which have been severely harassed by man or had their social order partially upset by antler damage have a delayed antler shedding. Fallow deer show this particularly. Nutrition also affects antler casting as does the age of the stag and the weight of the antlers carried.

Antler growth

Increasing day length and declining levels of the hormone testosterone start the growth of the cartilagenous substrate over which an outer skin of 'velvet' grows. This velvet is rich in blood and vascular tissue and is valued for medicinal purposes in some countries of the world. Once the antler substrate calcifies with the rise in testosterone level and decreasing day length, the velvet is shed and the male enters the 'hard antler' stage. When the hormone levels fall again, the bony areas at the base of the antler start to erode and the antlers are then shed from the pedicle[11].

In the first year of life, the pedicle is established and then the first small antler appears. These juvenile males are called 'spikers' (Reds) or 'prickets' (Fallow) and need to be carefully handled as these small sharp antlers can be quite dangerous. Dehorning spikers at an early stage of velvet growth can cause the growth of a larger set of antlers but harvesting this later-grown velvet may not be economic where quality is paramount.

Antlers have a number of functions:

- In the velvet they help to regulate body temperatures (this is debatable).
- They give off a scent after the velvet antler has been rubbed over the subcutaneous glands under the hind legs.
- Hard antlers are used as tools to dig for food in snow and to mark trees or threshing to mark territory.
- Hard antlers are used as weapons to compete for food where it is limited or in defence of territory and harem.
- Antlers enhance social status and are used as a means of recognition by other deer.

Stags take great care not to damage their antlers while in the velvet and any fighting that occurs is done by rearing up on the hind legs and striking with the forefeet. The loss of antlers, especially during the rut means an immediate loss of social status for the stag and he will be quickly challenged. In farming situations where all stags are de-antlered, the social order is not affected.

Castration

Castration of a male early in life (during the first year) will stop the development of the pedicle and hence no antlers will grow. These castrates without antlers are called 'haviers'. Naturally polled, fertile males are called 'hummels'. If more mature males are castrated, they will shed any antlers they have within a few weeks of the operation. Future antler growth will continue for about 4 months after castration. These antlers will stay permanently in velvet and will be misshapen.

Castrates can develop secondary sexual characters such as the enlarged neck, mane growth and deepening voice. They will follow females in oestrus and will mount them and show the flehmen response if treated with testosterone propionate.

The rut

Prior to the mating season or rut, Red deer are found separated into bachelor groups of males over 2 years of age, and female groups

with juvenile males settled on a home range. At the start of the rut
Red stags separate out and move onto rutting areas where they 'roar'
and show other threat displays while waiting for the hinds to come
into oestrus.

These rutting stands tend to be set up where several hind home-
range areas overlap[16] as shown in fig. 26. Fallow stags remain solitary
outside the rut.

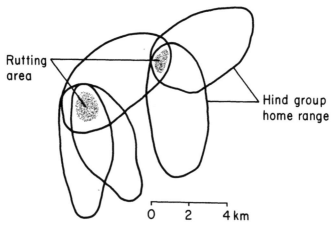

Figure 26 Location of rutting areas in relationship to normal hind home ranges[16]

A Red deer stag reaches puberty between 9 and 15 months of age
(Fallow 12–15 months) and is fertile by 16 months old. However, he
usually does not have the status to set up a harem before 5 years of
age, but this depends on competition from the older stags. During
the rut a Red stag's daily grazing time drops from about 65% of the
24 hours to less than 5% and his behaviour is divided among the
following activities. Note that these details are for Red deer as other
breeds differ.

Roaring
Undertaken in series at not more than 4 second intervals between
roars. The stag roars while carrying out challenges to other stags, at
about four bouts/minute. While roaring he goes round with neck
outstretched and the strength of his roar is generally related to his
fighting ability which prevents many encounters from starting. Stags
can go on mating for longer than the rut and they maintain their sex
drive for almost 6 months. Thereafter they are sexually quiet.

Herding hinds

The stag herds his harem using a chin-extended threat posture as he runs around the group. He smells the vulva of each hind about every 30 minutes and he also smells the ground where the hinds have been lying and after a flehmen response he investigates them further. He regularly chases the hinds, trying to mount them while flicking his tongue.

Observations in New Zealand show that wild Red deer stags attempt to defend harems of up to twenty hinds. On Rhum in Scotland, studies[15] showed harems to be six or more hinds and there was interchange between them. The largest harem was twenty-two hinds, and 80% of hinds were held in harem groups.

Young stags (from 2 to 7 years old) are unable to defend harems in competition with older dominant stags. The younger males are tolerated near the harems and they do occasionally sneak in and mate hinds when the dominant stag's attention is diverted.

Mating

The stag guards hinds coming into oestrus for up to 12 hours before they will stand. The odours excreted by the hind excite the stag and teeth grinding and tongue flicking are often seen. He pursues the hind and if she urinates he sniffs and gives the flehmen response. As the hind moves less and less the stag places his chin on her rump and he licks the vulva.

After mounting, the male stops for intromission and after a pelvic thrust he lunges forward during ejaculation with near-vertical body position and legs leaving the ground. The rapid lurch forward may lead to dismounting or injury.

After dismounting the stag roars repeatedly and urinates, stands with his head low and then begins to guard the hind again. Another mating may be attempted after half an hour.

Fighting

Antlers play an important part in group cohesion during the hard-horn season (8 months) as they allow a very stable dominance hierarchy to develop. Dominance is dictated by fighting prowess, so the older, heavier animals are the victors in fights. They also have had more fighting experience.

Threats are displayed regularly between stags before the rut by lowering the head and pointing the antlers to the opponent. This is usually enough to maintain relative dominance. If the animals are evenly matched, wrestling and a shoving contest may occur with antlers locked and little damage done.

In the rut, rival stags may have antler clashes interspersed with a threat display where the stags move on a parallel path to each other over short distances, where they display their lateral prowess, body size, mane and neck-roaring for periods of 3–15 minutes. This then erupts in head to head fights. Studies[14] show that the challenger always keeps an escape route open.

Much of the rut fighting is highly ritualised and little damage is done. Observations show that about 6% of rutting stags are injured each year in the wild usually being gored in the flank. There can be considerable fighting between stags after the rut when males are regrouping and before their antlers are shed.

Wallowing and thrashing bushes

A dominant stag will roll in a mud wallow, masturbate and urinate over the brisket and forelegs to advertise his presence to rivals. He thrashes at bushes and trees with his antlers to mark a territory and is most vigorous on the side of his territory nearest a rival stag.

The female

The reproductive behaviour of the female is dictated by the rut in the autumn and the birth and care of calves in the spring.

Red deer females reach puberty at around 16 months of age when they weigh about 65 kg. Body weight dictates reproductive success and hinds below 52–55 kg generally do not become pregnant. Small hinds conceive later, calve later and have greater calf mortality. However in the wild, hinds on home territory with good feeding have few reproductive problems.

The rutting areas are set out in hind territory and do not vary much from year to year. Females are attracted to these areas by the male roaring and they may drift from one area to another.

The Red hind comes into oestrus every 18 days (26 for Fallow) and will continue for about eight cycles if she does not become pregnant. Oestrus in Red hinds lasts about 24 hours during which time she may be served more than once by the same stag or by a number of different stags. Depending on the stage of the mating season and the physical condition of the stag, the mounting to ejaculation ratio may vary from 1:1 to 23:1 or more.

The hind in oestrus produces a secretion described by some as having a strong, penetrating smell and this combines with another sweet, musty odour. The stag's strong smell is also left on the hind after she is mated. Hinds in oestrus hold their tails high and there can be fighting and threats between them. The presence of last year's calf does not appear to affect behaviour.

Hinds may elicit stag interest by running past them with head low and neck extended and chewing. On one of these forays she suddenly stops to allow mating to occur. The hind urinates and this excites the male further once he has licked the urine and her vulva. (See reference 12.)

Hinds on Rhum in Scotland produce about twelve calves per life-time in the wild if well-fed but farmed deer will produce more. Gestation ranges from 228 to 231 days and Red male calves are carried longer than females. Gestation in Fallow is about 229 days. Males are heavier at birth, grow quicker and milk the dam out more effectively. As a result, dams suckling males are less likely to breed the next season and if they do they may calve late.

Red and Fallow hinds maintain their own matriarchal dominance orders which they keep throughout the year and this aids group cohesion.

Production

Deer farming

The advantages of deer over conventional farm animals are regularly described as their ability to thrive on second class land, good reproductive performance, good growth rates and feed conversion, lean carcasses with a high proportion of primal cuts and a range of valuable byproducts.

While deer can be farmed on poor land, they are not as hardy as often assumed and need good care and attention if they are to be profitable. Research[7] showed that the lowest critical temperature for deer is +5°C, whereas it is 0°C for sheep and −5°C for cattle.

There are now a number of texts on deer farming (e.g. reference 19). Behavioural work started with the study of F. Fraser Darling[4] on Red deer, and many years of Red deer studies on the island of Rhum in Scotland have been invaluable in documenting behaviour. Recent studies of farmed deer in New Zealand have built on this early work.

Capturing wild or feral deer

A number of methods have been used to capture deer:

- Find and capture calves near birth, then hand rear them.
- Trapping. There are a number of techniques[3,19] :

- Building a large fenced trap (20 m x 40 m) in which food is placed or grown and deer enter the trap by jumping into it, and from which they cannot jump back out. Live tame deer may be used in the traps as decoys.
- A large fenced trap with lead-in wings where on entry the deer trip a wire which closes a gate behind it. The gate may be closed by an observer.
- A walkover trap where deer can walk over a fence into a farm but cannot return to the wild.

- Shooting. Here deer are usually shot from a helicopter with a tranquilliser dart and a small radio beacon (from a double-barrelled gun). The deer is left and is located and collected when tranquillised.
- Bulldogging. Here a helicopter pilot flies near the deer and his crewman leaps out and grabs the deer, grounds it and ties its legs together. (See reference 19.)
- Netting. Again done from a helicopter where a net is fired from a gun to entangle the deer. It is then leapt on by the crewman. (See reference 19.)
- Mustering. Deer can be driven by helicopters or hunters with dogs into large, winged traps. With Fallow, long strips of coloured plastic can be spread through the forest and these moved nearer and nearer to the trap at intervals of a few days. Fallow will not go across these barriers. Toilet rolls have even been used as a cheap, short-term visual barrier festooned through the scrub.

Whichever method of capture is used, it is critical that the deer are treated with appropriate drugs for stress and that heart rate is monitored during this process. Veterinary advice should be obtained on these aspects.

Transport
Newly captured deer are lifted out after helicopter capture in a net or bag to prevent damage to neck muscles and ensure normal breathing. Blindfolding can be used to reduce fear and stress.

During transport farmed hinds and calves should not be mixed with stags and some farmers allow their deer to settle overnight in a darkened barn before transport the next day. Similarly they will hold animals in a darkened area before release. Farmed deer are regularly transported in closed-in and darkened trailers and crates. Good ventilation is critical especially when the vehicle stops.

Releasing
Great care is needed in releasing deer after capture as high mortality

may result through shock and stress. The release area should be large enough to allow the animals plenty of space, some trees to allow cover, and the deer should be left undisturbed for as long as possible.

Observations of Red deer show that their first reaction is to pace the fence lines and then go into cover and sulk. Water should be available and if possible a wallow. It is not wise to release them into groups of other deer as the newcomers will be harassed and this increases overall shock.

Red deer pace and run along fences and try to leap them from an oblique angle. Fallow run along fences and deep ruts develop which on slopes can cause erosion after heavy rain. Fallow run at fences head on and can force their way through the netting squares.

Captured deer vary in the rate at which they settle. Deer born on deer farms generally have fewer problems as the farm becomes their 'pad' and they exhibit a strong attachment to it. They can then be rotationally grazed around their farm.

Fencing

A 2 m high boundary fence is required if deer are to be adequately restrained and netting fences are preferable. These provide a reasonable visual barrier and this can be reinforced at pressure points by more wooden supports (battens) or foil as a glitter.

Barbed wire should not be used because of damage to stock. Before electric fences are used deer should be trained to respect them.

Figure 27 shows some of the features of fences in New Zealand. The electrified outrigger prevents damage by fence walking.

Plastic-covered netting is recommended along the bottom of the fence in races and approaching yards where pressure on the fence increases and small animals may escape. Close boarding can be used.

Rutting time is hard on fences, especially when antlered stags confront each other at night through the fence. De-antlering these stags will stop the damage and unbalance the stag for a while so he will not jump. However, de-antlering must be done without stress otherwise the stag's reproductive performance may be affected. It is not possible to run de-antlered and antlered stags together.

Stags in hard antler are a great danger to stockmen and each other when yarded and handled. The hard antler should be removed just above the coronet leaving no spikes. Many deer are injured by rubbish left behind by fencers. A concerted effort is necessary to collect up short lengths of high-tensile wire and other trash after the fence is completed. If deer are trained in paddocks which are well-fenced, they gain a respect for them. This respect can then be pre-

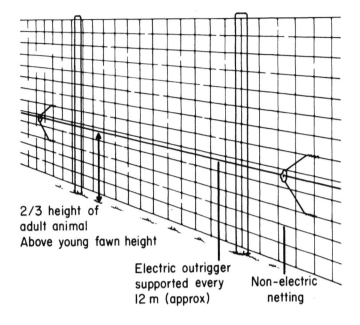

2/3 height of
adult animal
Above young fawn height

Electric outrigger
supported every
12 m (approx)

Non-electric
netting

(a)

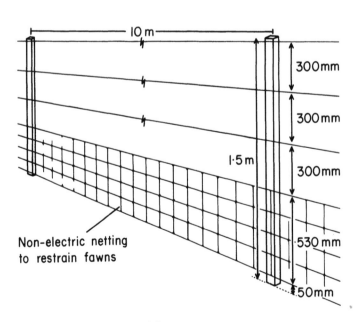

10 m

300mm

300mm

1·5 m

300mm

530 mm

50mm

Non-electric netting
to restrain fawns

(b)

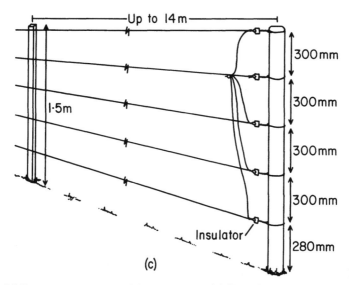

Figure 27 Fences recommended for Red deer: (a) Electric outrigger on standard fence; (b) combined electric and netting fence; (c) completely electrified (source: Gallagher Electronics, Hamilton, N.Z.)

served and built on by careful handling. Few fences will hold highly stressed animals.

Siting of fences should take into account the needs of deer for shade, shelter, water to drink and to wallow in. Wallows can be sources of disease and some farmers do not provide them. If possible gates should be sited in areas on tops of hills to which animals draw naturally when disturbed. Very wide gateways are advisable. Some farmers provide short lengths of collapsible fence at critical stock movement points.

Yards

There are many designs available for deer handling yards and these depend on the routine management to be done and the personal preferences of the owner. A number of general principles are presented for people considering yard construction and handling facilities:

- Walls should be solid, 2.2 m high for Red deer and 2.6 m high for Fallow.
- Yards should be covered for weather protection and at least a portion darkened.
- A squeeze crush can be used in combination with a head bail. A crush with sloping sides is often used for Fallow. Touching them

on the back will cause them to jump along the race with sloping sides. The floor can be dropped and the deer are then 'cradled'.

- Without a crush deer can be held in corners behind plywood sheets carried by assistants.
- Small pens to hold six to eight animals are preferable to single-file raceways in which they pile up. For Reds pens 1.2 m x 2.4 m are ideal. A short entry is needed from this holding pen to scales or squeeze crush.
- Circular, hexagonal or octagonal yards are preferred by some operators to avoid corners in which stock can be crushed and smothered. Fallow deer have been handled in a 0.9 m deep pool to avoid injury, but this was found to be impractical.
- Bolts and catches on doors between yards need special attention. These should be readily accessible from both sides, be solid enough to withstand pressure and open smoothly and easily for rapid escape by the operator. Gaps at the hinges must be avoided.
- Floors in pens. Opinion varies on the material used as some materials, e.g. earth or gravel can harbour footrot organisms. For yards used regularly clean sawdust on a firm earth base is useful. Concrete is generally too hard and slippery.
- Inspecting other handling facilities for different breeds of deer and obtaining handlers' opinions is very valuable. Retaining flexibility during the early use of the yards is also important so alterations can be made cheaply.

For further information see Further reading, p. 124.

Handling

The following points summarise current knowledge in handling Red and Fallow deer:

- Handlers should move slowly, positively and without hesitation among deer. Deer seem to sense a lack of confidence in handlers.
- Keep the number of people handling deer to a minimum. Quiet handling is essential.
- If dogs are used, they must be under complete obedience control. Deer move best if people and dogs are familiar to them.
- Deer move best along their established tracks. Leaving gates open overnight before mustering next day is good practice.
- Deer move and handle best in the morning. Farmers comment that they are more unsettled after noon.

- Deer should be moved through the handling facility a number of times to avoid them associating it with pain or stress. For example yarding can be done as part of moving between grazing areas, or they can be fed in the yards.
- Handlers should wear hard hats, face visors and protective clothing. Deer can react extremely quickly by standing on their back feet and boxing with razor-sharp front hooves. The normal and boxing postures are shown in fig. 28. Fallow deer also jump high from a standing start and can land on the head or shoulders of the handlers.

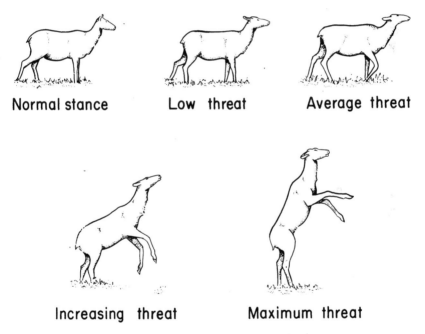

Normal stance   **Low threat**   **Average threat**

Increasing threat   **Maximum threat**

Figure 28 Threat postures of Red deer (source: Ruakura deer unit)

- Handlers need to recognise the threat signals of deer as follows: Striking with forefeet – head held high and preorbital gland wide open. Biting – head tilted, upper lips raised, tongue protruded, grinding teeth, whites of eyes showing and hissing.

These comments refer mainly to Red deer. Some stockmen keep a riot shield handy to protect themselves or may erect a 'dodging post' in the yard to hide behind if attacked, especially by a stag.

Removing antlers – velvetting

In removing antlers for velvet, to achieve highest market return the following points are important:

- Check the local regulations regarding harvesting velvet.
- Animals should be yarded and handled in small batches to avoid bruising the velvet.
- Harvesting should take place at the correct stage of growth, which is usually 55–60 days post-casting for 2–3 year olds and 60–70 days for mature stags.
- Veterinary assistance is needed for the approval and administration of drugs to tranquillise the animal and to apply tourniquets to the animal's pedicle. These are released 5–10 minutes after cutting. Animals that are used to being handled and are familiar with the yard environment take less time to recover from tranquillising.
- After sawing off level above the pedicle, the antlers should be immediately inverted to retain the blood supply, placed in a plastic bag, cooled and frozen. Average production is 1.8 kg of velvet antler from Red deer stags but top yields of 3.5 to 5.0 kg are achievable.

A stag's potential for velvet production cannot be predicted accurately from his production as a spiker. It is necessary to wait until his next year's production to predict lifetime potential. Adjustment tables are available for New Zealand conditions[10].

Regardless of whether antlers are harvested for velvet, they should be removed for the safety of stockmen and others who may need to handle stags. Some breeders are loath to remove the antlers from their top stud sire so that they can demonstrate his antler potential to buyers of his progeny. Extreme care is needed in handling such dominant animals. Remember that deer react like lightning whether they are goring, kicking, striking or biting and vigilance is paramount in close handling.

Farm mating

In a farming situation younger stags can be used than in the wild, especially if given single-sire groups. Mating ability on the farm can be independent of fighting prowess. The main points in mating farmed deer are these:

- Body weight. Stags should be well fed to ensure a good body and physical fitness coming up to the rut.
- Rut areas. Although these can be of any size, dominant stags in small areas can hold more hinds than they can successfully mate. Single-sire mating groups can be run in small areas. In larger herds run in bigger areas for mating by a number of stags, several harems will be formed. Area as such has not a major effect. The feed available in the area is important in dictating management and shifting of the stock.
- Mixing stags for mating groups. Stags should be mixed prior to antler growth if possible. If the dominant stag is becoming exhausted, introducing a new stag may not break up his harem. Overworked stags should be taken out from groups to allow other stags to mate. Subordinate stags search out females in heat but most of the mating is done by the dominant stag.
- Studies[2] concluded that one stag to fifty hinds would be an effective joining rate for genetic improvement and replacing a single sire by another every 3 weeks would ensure a high conception rate.
- Swapping stags during the rut can be a difficult management procedure and great care is needed if stags have antlers.

Supplementary feeding

Feeding supplements to deer, especially newly captured animals, greatly assists their taming and handling. Winter is an ideal time to establish these feeding and management routines. Feed can be used to attract shy animals into new areas or handling yards. Mineral or salt licks can be used similarly.

Farmed deer have similar food preferences to cattle and sheep although in the wild they are mainly browsers. They soon learn to relish pelleted concentrate feeds and these are useful as handling aids when the deer need to be attracted[8].

Slaughter and shooting

Slaughtering farmed deer for venison production, especially if the meat is for export to other countries (as from New Zealand) is now done in approved deer slaughter premises using similar slaughter methods to cattle and sheep, and packed at special game packing houses. When deer are shot in the field the rifle and ammunition used should be suitable for the species of deer being hunted. Soft-nosed or hollow-pointed ammunition is generally recommended. Rifles of less than 6 mm calibre have limitations and should only be used on deer by expert marksmen who can always place their shot precisely for a clean kill. Rim fire 0.22 calibre rifles should not be used on deer or other large game.

Welfare

From the limited information currently available on deer behaviour, a number of recommendations can be made to meet the welfare needs of Red and Fallow deer. These are:

- Deer need shelter and shade and cover for the young.
- They need to be fed to meet their nutritional needs.
- As they suffer from similar diseases to other farm stock, a planned animal health programme is essential.
- Deer need clean water to drink and Red deer like to wallow. This latter must be balanced against the potential of wallows in disease spread.
- Proper handling facilities are essential to reduce stress in the stock and prevent injury to both deer and stockmen.
- Fences must be of the correct height, well maintained and visible. Pressure points on the fences must be reinforced and made especially visible. Animals should be trained to respect fences (especially electrified ones).
- Yards. Designs should prevent injury to animals and handlers and be planned for throughput with minimal stress.

Concern over the welfare of farmed deer is mainly focussed on the removal of velvet antler. In the minds of many people the use of the velvet antler in Eastern medicines does not justify the stress caused by the practice or the indignity inflicted on the stag. There is no question that antler removal, whether it be for velvet or for safety of stockmen, should be carried out under veterinary supervision using the most modern procedures.

Concern is also expressed over the stress and indignity the animals are subjected to during capture and retrieval. Again modern operators are highly skilled in this practice and the exercise is only worthwhile if losses and stressed animals are avoided. Veterinary approval of the drugs and methods used is important.

Some people are concerned with the stress and suffering associated with slaughter. A deer correctly shot out in the open probably suffers least stress of all but wounded animals suffer badly and may die a slow, lingering death, some never to be found. When slaughtered in approved abattoirs there is some stress but it is little different to that experienced by other farm stock.

References

1. Arman, P. (1974) 'Parturition and lactation in Red deer.' *Deer* **3**(4), 222–3.
2. Bray, A. (1978) 'Observations on behaviour at mating in Red deer herds: summary.' (Mimeographed.)
3. Brookes, R.A. (1977) 'The capture and farming of Red deer from the wild.' *N.Z. Agric. Sci.* **11**(4), 162–4.
4. Darling, F.F. (1963) *A Herd of Red deer.* Oxford University Press, 213 pp.
5. Espmark, Y., and Langvatn, R. (1979) 'Cardiac responses in alarmed Red deer calves.' *Behav. Proc.* **4**, 179–86.
6. Fennessy, P.F., Moore, G.H. and Muir, P.D. (1981) 'Artificial rearing of Red deer calves.' *N.Z. J. Exp. Agric.* **9**, 17–21.
7. Fletcher, T.J. (1980) 'Farming deer.' *Appl. Anim. Ethol.* **6**, 87–8 (abstr.).
8. Ford, G.A. (1977) 'Red deer on a hill country sheep and cattle farm.' *N.Z. Agric Sci.* **11**(4), 168–9.
9. Gilbert, B.K. (1974) 'The influence of foster rearing on adult social behaviour in Fallow deer (*Dama dama*).' Paper 11 in *The Behaviour of Ungulates and Its Relation to Management*, I.U.C.N., Morges, Switzerland, pp. 247–73.
10. Giles, K. (1980–81) 'Deer management.' *Deer Fmr.* Summer issue, pp. 29–30.
11. Goss, R.J. (1968) 'Inhibition of growth and shedding of antlers by sex hormones.' *Nature*, **220**, 83–5.
12. Guinness, F.E., Albon, S.D. and Clutton-Brock, T.H. (1978) 'Factors affecting reproduction in Red deer (*Cervus elaphus*) hinds on Rhum.' *J. Reprod. Fert.* **54**, 325–34.
13. Kelly, R.W. and Drew, K.R. (1976) 'Shelter-seeking and sucking behaviour of the Red deer calf (*Cervus elaphus*) in a farmed situation.' *Appl. Anim. Ethol.* **2**, 101–11.
14. Lewin, R. (1978) 'Rutting on Rhum.' *New Scient.* **80**(1129), 540–2.
15. Lincoln, G.A. and Guinness, F.E. (1977) 'Sexual selection in a herd of Red deer.' In *Reproduction and Evolution*, eds. J.H. Calaby and C.H. Tyndal-Briscoe, Australian Academy of Science, pp. 33–8.
16. Lincoln, G.A., Youngton, R.W. and Short, R.V. (1970) 'The social and sexual behaviour of the Red deer stag.' *J. Reprod. Fert., suppl.* **11**, 71–103.
17. Moore, G.H. and Cowie, G.C. (1980/81) 'Hand-rearing deer calves.' *The Deer Fmr.* Summer issue, pp. 27–8.
18. Topinski, P. (1975) 'Abnormal antler cycles in deer as a result of stress-inducing factors.' *Acta. Theriol.* **20/21**, 267–79.
19. Yerex, D. (1982) *The Farming of Deer* ed. D. Yerex, Agric. Promotion Assoc. Ltd, Wellington, 120 pp.

Further reading

Anderson, R. (1978) *Gold on Four Feet*, Ronald Anderson Associates Pty., 160pp.

Bannerman, M.M. and Blaxter, K.L. (1969) 'The husbandry of Red deer.' *Proc. Conf. Rowett Res. Inst.*, 79pp.

Cadman, W.A. (1966) *The Fallow Deer.* Forestry Comm. Leaflet, HMSO Leaflet No. 52.

Chaplin, R.E. (1977) *Deer*, Blandford Press, 218pp.

Chapman, D. and Chapman, N. (1975) *Fallow Deer*, Terrance, Dalton Ltd, Suffolk, 255pp.

Fraser, A.F. (1968) *Reproductive Behaviour in Ungulates*, Academic Press, London, 202pp.

Young, C.D. (1979) *Public*. No. 50. Scot. Agric. Colls. 26pp.

6 Horse

Over the centuries, the size, speed and grace of the free-ranging horse has always been admired by man and has presented him with a regular challenge. The horse has been used for transport, traction, war, sport, companionship and food but despite this, studies of its behaviour are few and recent.

Modern interest in the horse centres around the racing industry and riding for recreation and sport, as well as a renewed interest in the work horse due to the rising cost of oil-based energy. There is a growing interest by conservationists and geneticists in the populations of wild and feral horses throughout the world. Also people concerned with welfare and cruelty know that the horse is not exempt from maltreatment.

Behavioural definition

The horse is a large, grass-eating non-ruminant animal that inhabits drier rangelands, moving up to 16 km a day from grazing to water. Its home range area varies widely depending on food supply but can be as large as 1000 ha.

Horses live in various groups or bands, the most frequent being a family band or harem of one stallion, one or several mares and their offspring. Bachelor bands are commonly made up of a solitary male or up to eight males in association. Horses in groups develop strong social relationships.

Horses communicate by a variety of vocal calls, different body postures involving tail, ears, mouth, head and neck as well as odours associated with dung rituals. They have well developed senses of smell, touch and hearing and escape predators by speed and use bucking, jumping and kicking as defences. As a harem-forming species, the stallion covers his mares but not his fillies keeping them under surveillance, especially during oestrus. Mares generally foal in spring and summer and are usually mated a few days after foaling. Foals are precocial and can follow their dams a few hours after birth. As a follower-type species, the foal is protected by the mare and learns the trails, feeding patterns and fear of predators or man from its mother. Horses are large, active and curious animals and they respond well to stimulation.

Birth

Mare's behaviour before birth

Mares tend to become lethargic as they near term and some may become aggressive. Gestation varies widely from 305 to 365 days with a mean of about 344 days. Spring-born foals are carried about 5 days longer than those born in summer, with males carried about 2 to 3 days longer than female foals.

Signs of birth in the mare are swelling of the vulva, udder and teats from which milk wax may exude about 2 days before birth. About 4 hours before foaling there may be sweating above the elbows and on the flanks and body temperature drops 5–6 hours before birth.

Mare's behaviour during birth

Labour is usually in three stages. The first is where the mare stops feeding, becomes restless and walks in circles looking back towards her flanks. She kicks and starts repeatedly lying down and getting up again, slapping her tail against the perineum, crouching, standing in a straddle position and urinating. Often some milk dribbles from the udder, the head is lifted high and the upper lip retracted[1]. When the water bag (allanto chorion) ruptures labour starts in earnest. This stage varies in length, but usually lasts about 4 hours.

If trouble-free, the second stage of labour in the mare is very rapid and is usually over in about 15 minutes (range 10–70 minutes) and some mares may then rest for periods up to 30 minutes. Once the birth sac appears the foal's front feet are followed by the nose and head which requires extra straining to expel. Once the head is free the rest of the body quickly follows and the foal's back legs are freed when the mare rises. Rupture of the birth sac is usually accomplished through vigorous movements of the foal's head and long legs. During labour in stalls mares tend to lie with their backs close to the wall. The foal is born when the mare is lying down but the mare will stand up if disturbed.

The third and final stage is where the placental membranes are expelled and this usually takes up to an hour. The mare does not normally eat these.

About 80% of mares foal at night peaking around midnight[18]. One observer[20] maintains that mares give birth only when they feel comfortable, hence the wide variation in gestation. Night foaling certainly gives extra work to horse owners if they wish to be present at the birth. Foaling has not been observed in feral studies.

As less than 2% of mares will experience difficult births (8% in thoroughbreds) there is little need for man's interference. X-ray studies indicate that the normal presentation protects the foal's rib cage during birth. Dystocia in thoroughbreds is commonly caused by fused joints in the forelimbs. Relief of malpresentation may result in limb deformity[18].

The mare's reaction to the foal

Care-giving behaviour by the mare is shown from the time of birth and seems triggered in part by the taste and smell of the allantoic fluid. Mares sniff the ground where these fluids have fallen and give a lip-curl response. Sometimes they appear to be seeking a foal before labour is completed.

The mare generally spends several hours licking the foal starting at its head. Some younger mares may be excited by the foal and squeal and strike at it with their front feet until it stands up. The foal may also be kicked while it is searching along the mare's belly for the teat.

The foal approaches the mare's udder from her head and parallel to her body. Good, experienced mares stand still during this time and nuzzle the foal's rear. The foal suckles with neck outstretched and close in to the mare's flank to avoid being kicked. The mare keeps the foal in close physical contact with her from birth even when moving. Feral stallions will round up straying foals.

The foal's reaction to the mare

Once the foal is born it starts to move by swinging its forelegs and within 15 minutes uses its front feet as levers to hold its sternum off the ground. The hind legs start to flex at about 45 minutes after birth when the foal attempts to stand. Stimulated by the mare's licking, physical contact and nickering, the foal starts teat seeking.

The foal is able to suck as soon as 30 minutes after birth but this depends on the movements and behaviour of the dam. Observations show that the average time to suckling is 111 minutes. Suckling is frequent and studies have recorded fifty to seventy-five/24 hour period during the early weeks of life. Foals need to suckle the mare's colostrum within 8 hours of birth. Once a foal has suckled several times the mare fully accepts it and it can easily locate the teats. Suckling lasts from 15 seconds to 2 minutes and is initiated by the foal. Observations show that 75% of sucklings are terminated by the foal. Meconium is normally passed within 72 hours of birth.

Once suckling starts the foal's tail is active. A characteristic pose is taken with the legs spread and tail lifted 40% higher than usual. The first attempts of the foal to lie down result in collapse but correct posturing for recumbency is soon perfected. A foal can orient its head, use its eyes by 25 minutes and move its ears by 45 minutes from birth. The vocalisation levels of the foal increase until 2 hours after birth which is then quite a noisy time in the horse's life. Sleep soon begins in the second hour of life and foals sleep lying down[22].

Foals resist restraint from a few minutes of age but if handled by man in the first hour of life will habituate better and be more self-confident when older. Handling by humans will not upset the mare's response to the foal[23]. By 12 hours of age, a foal can nip its side to displace insects, move its tail and legs, urinate, walk, trot, gallop, play and nibble grass. While moving with their dam foals often stop abruptly and back into her. This is intended to stop the mare so that she will mother the foal. As follower species, a feature of the horse is the transfer of habits from mare to foal (fig. 29).

Rearing

Foals reared in small groups develop better than those reared by themselves. Social play is important in the proper development of the foal and determines later behaviour. Research information is scarce in this important area.

Foals reared on machines show aggressive snapping behaviour

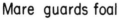
Mare guards foal   Foal associates with mare

Figure 29 Mares and foals keep together and the foal learns many habits from their dams

towards older horses, and they do not learn fear of objects normally learned from their dams. Colt foals frequently mount others in play when a few weeks old but fillies do this less frequently.

The mare does not necessarily know her own foal right from birth and other foals can be fostered on to mares for up to 3–4 days after birth. It seems that recognition is largely based on smell and taste. Mares will leave their own dead foals after a day or so. Bonding is firmly established in 2 hours and the foal soon learns to avoid objects and people through its dam's nickering.

Weaning takes place naturally about 9 months of age but in commercial studs, foals are weaned at about 5 months old.

Reproduction

The male

Although males can produce an erection of the penis at 2–3 months of age, they can only achieve successful mating by about 15 months old.

In the wild, the stallion separates the mare several days before ovulation keeping her under surveillance to determine when standing oestrus occurs. The harem stallion keeps other stallions away from the mare during this time and may fight to protect her. All this protection and courtship helps to stimulate the male. In horse studs with hand mating this stimulation may not occur.

The stallion's first approach is to sniff, lick or nuzzle the mare's perianal area. The mare urinates and stimulates the stallion to give a flehmen response. Considerable vocalisations occur from both partners before the stallion mounts. He may move his head across the mare's rump or nip her hocks and toss his mane. After mounting, intromission occurs within about 10 seconds. Studies[21] have shown that at least seven thrusts are needed to bring about ejaculation. These are quite rapid, over about 20 seconds. The ejaculate is exuded in five to ten bursts with 80% of the sperm present in the first three jets. During ejaculation, pelvic thrusting ceases but tail flagging continues. There is a refractory period by the male after dismounting of about 20 minutes. The mare in oestrus still continues to display. The libido of stallions varies throughout the year and stallions can copulate three times or more per day with ease.

Mating problems

A stallion must have a full erection before intromission is possible and courtship is important to stimulate this. Some stallions show sexual interest but will not mount while some show no interest at all. Some stallions will only mount when a specific horse is present and some are vicious and can cause injury to the mare and handlers. Even when gelded, these vicious animals are still dangerous. Normally a stallion's sexual activity is only limited by season of the year or physical exhaustion.

Some of the behavioural reasons for a stallion's inability to serve a mare are these:

- A frightening or painful early experience with man or other horses.
- Stallions fitted with mating rings to prevent them having erections at shows and exhibitions.
- Falling on slippery floors or pain suffered during mating such as severe kicking from a mare.
- Banging head on rafters when mating indoors.
- Masturbation reducing sexual drive.
- Over-use as a 2-year-old.
- Pain in feet, limbs or joints when mounting.
- Excessive discipline given by inexperienced and frightened grooms.

Veterinary advice should be obtained with these problems. Changing the stallion's environment, his handlers and letting him mate away from other males may help.

In horse studs, stallions are usually led up to a mare with a gate or kicking board between them, or the stallion may be led along a line of mares held in a narrow pen. Sometimes the mares are hobbled, held by a lip twitch or have large felt boots fitted to their back feet. Some may be blindfolded. All of these devices are to get the mare to stand quietly and prevent her kicking a valuable stallion. It is little wonder that reproductive performance in these situations is often poor. Anxiety, apprehension and restraint inevitably add stress to all concerned.

Teasers

Male teasers are used to test whether a mare is in standing oestrus to save the stud stallion's energy and prevent him from injury. An entire male may be used as a teaser but is usually much smaller than the mare so that copulation is physically difficult. The sexual response of mares to teasers of different colours, contrasting size and vocal repertoire is quite unpredictable. Two of the possible teasing systems are shown in fig. 30.

A number of studies have reported the use of vasectomised teasers

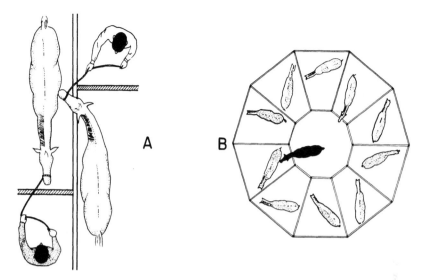

Figure 30 Facilities used on two farms for testing mares for oestrus. A, race with kick board; B, teasing mill[15]

fitted with a raddle which marks the mare when the male comes down on her. No mark was left in one out of every five matings and one in every three mountings. However, a mare with a 4-day or longer oestrus was always marked. Most of the sexual activity was seen in the early morning or late evening. Using a teaser allows a natural harem situation to develop in the field and mares identified as being in standing oestrus are brought in for entire mating. Regular health checks are necessary with vasectomised teasers in case they become a source of venereal disease spread.

The female
The female reaches puberty at about 12–24 months of age. Her breeding season is initiated by an increasing daylight pattern. The mare's cycle length averages 20 days and oestrus averages 5.7 days with a wide variation from 1 to 24 days. Oestrus shows a seasonal variation, e.g. 7.6 days in spring and 4.8 days in summer. Most mares ovulate within 48 hours of the end of oestrus and ovulation occurs mainly at night[7]. Oestrus without ovulation is rare but ovulation without oestrus is more common. The mare displays a very wide range of responses during oestrus.

Four phases of pre-coital response have been outlined[7]. These are:

- Greeting, including nosing and face-nipping.
- Active interchange including biting and vocalisation.

- Oestrus displays like posturing, tail arching, clitoral flashing (winking) and mucous exudation. These are signs which attract and excite the male.
- Passivity when the mare stands still and orients towards the stallion.

Careful documentation[7] of observations has now produced a definition of oestrus in the horse as follows:

A mare is in oestrus when she stands firmly with tail up while being mounted when one of the following signs are also shown:

- Urinating at any time during teasing.
- Winking the clitoris at any time during teasing or
- Tail raising before and after mounting.

Mares vary widely in their expression of oestrus and they can be divided into those displaying continuous or overt oestrus, split oestrus and covert oestrus.

This range of expression of oestrus would seem to increase mating success in harem species. Generally, the mare does not show overt oestrus behaviour until a stallion is present, though some stables now use a tape recording of stallion calls. Space is important to allow horses to move around and show oestrus signs. It also reduces the risks of injury to both horses and handlers by being crushed or kicked. At the first heat after birth (the foal heat), the mare may need to have her foal present or she may become upset.

Social behaviour

The nature of a horse's adult social life grows out of the early mare–foal socialisation and the amount of play and social contact with other horses when young. The main aspects of social behaviour in horses are described below.

Physical contact

When horses make contact with each other, they first meet nose to nose (naso–naso). Figure 31 shows the other points in contact when strange horses meet.

Vocalisation

Investigations have grouped vocal calls of horses in different ways. Here, seven basic calls are described[24], four vocal and three non-vocal and are summarised in fig. 32.

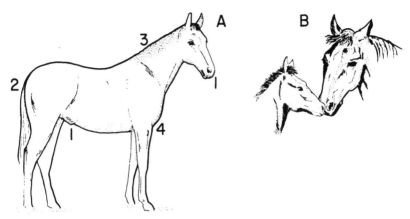

Figure 31 The points of body contact when horses meet. A, contact points between strangers; 1 = high, 4 = low; B, mare–foal contact (after Houpt (1977) – see Further reading, p. 149)

- Squeals – last 80–1720 msec.; are high-pitched harmonics ranging in loudness; made during threats and encounters between individuals.
- Nickers – last 250–1720 msec.; low-pitched and broad band uttered in a series of pulsations; occur prior to feeding in the range 6000 Hz; made by stallion during mating and between mare and foal at below 1000 Hz.

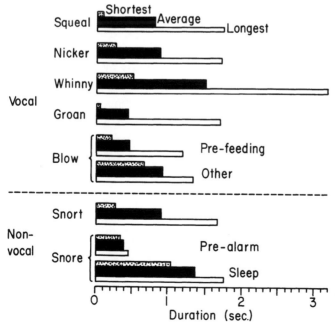

Figure 32 The duration and range of vocal calls in the horse[24]

- Whinnies − begin as squeals in harmonics and terminate in nicker-like bands lasting 500−3180 msec.; very audible and variable and generally of high pitch; occur between horses that are distressed or seeking social contact such as at weaning.
- Groans − last 60−1690 msec.; are broad band non-voiced sounds which lead into voiced sound; travel up to 10 m as heard by humans; occur during discomfort or anguish.
- Blows − last 650−1330 msec.; non-vocal sounds made as air passes through the nostrils; generally below 3200 Hz; they carry 30 m and occur during pre-feeding or in alarm.
- Snorts − last 280−1680 msec.; produced by blowing air from the nostril during nasal irritation, conflict or relief.
- Snores − non-vocal produced when air is inhaled; last 340−460 msec. in short snores or 1040−1750 msec. in long snores; have little part in communication.

Body postures

Body postures in the horse are used as means of non-verbal communication and clearly have advantages in survival against predators. Slight muscular movements of ears, tail, nose and lips are readily detectable by other horses. When travelling in single file, slight movements of the head indicate changes in direction to those following, ears back mean beware and a tail swish usually precedes a kick.

Rivalry between bands of horses in the wild further illustrates posturing. When stallions meet before an encounter, they move through a head-bowing sequence followed by prancing with curved front legs and stomping the ground with the front feet. Studies[2] showed that on 80% of occasions the encounter ended without further conflict. Head bobbing is also used by stallions to move or herd mares and foals. The tail squeezed down firmly between the legs is a clear sign of submission in the horse.

Dunging and urinating rituals

Horses are notorious for over-grazing certain areas and concentrating their dunging and urinating in other areas which they do not graze. Hence dung piles build up.

Dunging and urinating on the dung or urine of other horses is an important means of social communication. In feral horses, although dung is not used to mark out a territory or home range, it is used to indicate to other horses using that range the whereabouts and movement pattern of the stallion and his band. Horses, especially stallions, regularly sniff dung piles presumably to obtain information about other horses in the area. Usually the most dominant male is the last to mark a dung pile so from observations, it is possible to obtain a social rank among males.

Studies[5] of range behaviour in horses showed that from 120 dung piles examined, 57% were in current use, 45% were by regular trails and the density of piles was greatest near the water hole. Most piles (60%) were single piles, 13% were two piles side by side and then the numbers decreased up to five or six piles together. When piles got too high others were started. Studies have shown that 90% of dunging by males was in response to piles of dung they had investigated. During inter-stallion rivalry both males smell the dung pile, use it, return to it and repeat the behaviour. During a fight, defecation can occur up to three times.

On farms, mares tend to dung in a very specific area. Stallions will go to this area, investigate the dung and then direct their own elimination on the same spot. Stallions regularly urinate on spots where mares urinate. The result of this is that the area fouled by mares increases in size while the male area stays more circumscribed. This makes pasture management difficult as horses unlike sheep avoid grazing near their own dung.

Horses generally eliminate every 3–4 hours. They defecate while moving with tail lifted and deflected but they always urinate standing still after taking up a posture with hind legs separated and slightly crouched. The mare usually shows clitoral flashing after urination. Foals have been recorded to urinate up to twenty times/day in the first 2 weeks of life.

Horses can be encouraged or trained to urinate in a stable or box by freeing them from restraint, forking or whisking the straw bedding and whistling a long vibrating-type note. Urine can then be collected in a large stainless steel pan if needed for examination.

Fighting

Most fighting is seen among males. Most encounters are restricted to posturing but when real fights occur the contestants rear on their hind legs, attempt to neck-force the other horse down, kick at each other with their front feet and bite while rearing. The fight is usually terminated by the dominant male chasing the subordinate and kicking out at him with both back feet.

On farms, plenty of space must be given to horses to escape when harassed to avoid injury. The two main submission responses in horses are first the simple move away when threatened and the second is called teeth clapping. This latter is shown by a subordinate stallion after being beaten in a fight or when a stallion returns to the harem after a mare has refused to be mated by him.

Grooming and body care

Horses like to roll in soft earth and scratch their own bodies with their hooves, teeth and lips. A healthy horse enjoys rolling after exercise and especially when hot and sweaty.

When two horses come together they hold their ears high and outwards. They smell noses and necks and then start mutual grooming each other's necks and withers, working towards the tail while standing in a head-to-tail position. This behaviour is mainly nibbling and may last up to 3 minutes between horses of the same group or harem.

Sleep

Horses require sleep and although most of it is done while standing, they do sleep while lying down. Sleep in horses is classified as slow wave sleep (SWS), paradoxical sleep (PS) and rapid eye movement sleep (REM).

In humans, the PS phase includes dreaming and horses also twitch and whinny during PS. They drowse during the day but truly sleep at night. Their sleep cycle is short, e.g. 5 minutes SWS; 5 minutes PS; 5 minutes SWS, then awake for 45 minutes. Total sleep is about 3 hours/night but PS occurs only while lying down.

A horse if forced to stand continuously will be sleep-deprived. As a horse will not lie down unless it feels secure, it may deprive itself of sleep. Examples of situations which prevent horses from sleeping properly are:

- Hospitals where slow healing of wounds is noticed.
- Horses put in new strange paddocks, especially on their own may not sleep for 2 days.
- Horses stalled near human habitation where noise is sporadic and strange to them.

Being social animals, they relax more in a group. If some horses lie down, others follow and studies[11] show that dominant animals lie down first. Young animals (foals and yearlings) need more sleep than older animals and mares usually stand over their sleeping foals to protect them. Horses prefer to lie on dry earth or sand.

General behaviour

Feral studies

A number of studies of feral horses have now confirmed that harem bands are made up of a stallion and several mares and their offspring. Males which cannot maintain a harem associate in bachelor groups.

In studies of social dominance[8,14] the order of priority of reactions was:

- Complete kick with one or two hind legs.
- An intention to bite with a head movement and ears laid back and teeth bared, ranging to complete vicious bites.
- Head or neck bumps on another horse.
- Slow avoidance reactions.

Bites were 75% of total reactions and are clearly the most common aggressive act in the horse. Over 30% of all bites went into the flesh. Head bumps were 8%, leg kicks 6% and passive withdrawal 10%. Because biting is the most common aggressive response in the horse, handlers should be especially vigilant to avoid injury.

Movement and leadership

In studies of feral horses[5] stallions initiated movement of the band in 67% of cases, while the mares were solely responsible for movement in 23%. This herding together of the group by the stallion was done quickly by lowering the head and moving it side to side (called snaking). A mare from a strange group was occasionally herded in. A band may travel up to 16 km a day for watering. The stallion controlled the movement of his mares until other bands drank. He then led them to the water where in 94% of observations he was the first to drink.

Watering behaviour

Feral horses generally go to water once a day but not at a regular time. Fighting often takes place in the vicinity of water which is usually a limited resource. Elimination is avoided near water. Most of these feral horses do not move more than 6 km from water and they often roll near the water hole.

Working horses are usually watered after each spell of work. Heated horses should only be given limited fresh clean water as this routine avoids bouts of colic. Horses suck up water when drinking and up to half of their nostrils may be submerged. The information on water needs of horses is reviewed in reference 10.

Feeding

Horses spend long periods feeding averaging about 12 hours/day. When feed is short they may graze for up to 18 hours a day. Horses are selective grazers and will eat a wide range of pasture plants, herbs, shrubs and weeds. However in starvation conditions they will even eat mud and old dung. When pasture is very short, ingestion of soil can cause excessive teeth wear. The horse's digestive system is designed for small but frequent intakes of herbage, but excess amounts of legumes are not suitable for horses.

The normal habit of cropping grass is aided by the mobile and very sensitive upper lip. The upper lip forces grass into the mouth and the incisor teeth nip it off. After a few bites horses move to another area and prefer to graze facing up hill.

Foals look awkward when nibbling grass and when very young they compensate for their long legs and short necks by spreading their forelegs astride. As they become older, one foreleg is advanced while the other acts as a prop beneath the thorax or chest. Foals commonly eat faeces within 10 days of birth, presumably to initiate cellulose digestion in the colon.

Stall-fed horses become very accustomed to regular feeding routine and get upset if this is disrupted. Little time is spent eating concentrate feeds and unless sufficient roughage is provided, boredom will develop resulting in behaviour patterns classed as vices.

Learning, training and gentling

Horse senses

The first step in using behavioural knowledge in training a horse is to appreciate its sensory capacities. These are sight, hearing, smell, taste and touch.

Sight

The eyes of a horse are perfectly placed for cropping grass for half of its total life. It must raise or lower its head to focus on objects at various distances away from it, which it can do very quickly. The many rods in the retina allow it to see the slightest movement and colour can be perceived, starting from yellow, green, blue and red in that order. Horses need time to adjust their vision between light and dark, e.g. when moving from bright light into a dark box or trailer.

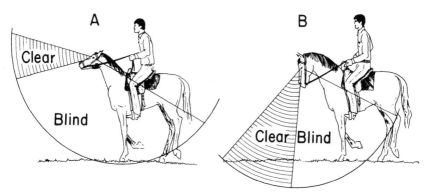

Figure 33 The head position and vision in ridden horses. A, head high; B, head low (after Miller (1975) – see Further reading, p. 149)

Only a limited area in front of a horse can be seen by binocular vision and the blind spot behind its head increases when the head is lifted as shown in fig. 33. Riders approaching a jump free the reins to allow the horse to lower its head sufficiently and use its limited binocular vision.

Hearing

Horses hear over a broader range than humans and their hearing is very acute in the high and low frequencies. For example, they can hear the approach of a storm before it is audible by humans. They keep alert at all times except in deep sleep and can swivel their ears 180°. Blocked ear passages cause great stress to a horse. The ears are valuable indicators to handlers of the horse's mood.

Smell

This is well-developed in the horse and wild horses can only be stalked up-wind. Strange horses smell each other when meeting naso–naso. They are very sensitive to smells in their environment such as dung, dirty feed troughs, musty feeds, bad water and certain plants.

Taste

Horses are attracted by sweetness such as sugar, molasses, water melon rind, apples and peaches, soda and beer. Foals have been shown in choice tests to prefer sweet foods and reject salty, sour and bitter tastes at about the same levels of acceptance as humans.

Touch

Touch is one of the most acutely developed senses in the horse and plays a major role in social life. An experienced rider uses pressure of knees, heels, spurs, weight distribution and general body relaxation to convey instructions to a horse.

General

The following are some likes and dislikes of the horse and general behavioural points which should be appreciated in their management and handling:

- Horses like a separate space for eating, drinking, sleeping and elimination.
- They do not like to lose their balance and fall, e.g. narrow bridges or slippery ground can frighten them.
- Horses in the wild usually walk, they trot in play and gallop only in flight, but without a weight on their backs. They never pull anything.
- Horses view with suspicion any new object met and have a good memory for previously frightening experiences.
- Familiar surroundings give them security. Terrified horses have been known to run back into their burning stables.
- Horses are curious animals and need stimulation to prevent boredom and stress.

Learning ability

A number of attempts have been made to test the ability of a horse to learn. This would have many advantages in saving time and money in training. Dixon[3] trained a horse to discriminate between twenty pairs of visual signs and the animal remembered these with ease. Other workers have used 'Y' mazes but found only a small relationship between learning and emotional temperament. The Hebb-Williams detour maze[13] was shown to be impractical on a large scale and a reversal learning task (where after animals have learned to discriminate between objects, the objects were switched) also has limitations. Research workers[17] found a difference between quarter horses and thoroughbreds in learning a visual task (5.4. versus 8.4 respectively) but most efforts to classify horses quickly and easily into good and hopeless cases have not been successful. Investigators are still trying.

Principles of training
The basic rules for working with horses[9,16,25] are similar to those already outlined for dogs and sheep. The main points are these:

- Do not scare the horse because this will set up flight responses.
- Remember punishment only suppresses behaviour, it does not eliminate it.
- Punishment incorrectly given, or given to excess makes behaviour worse as the horse's sensitivity is reduced and not improved.
- Reward training is the simplest and best method, but reward must follow the correct and required action as soon as possible.
- Rewards may include a pat on the neck, rest, food or being able to see new and interesting sights.
- Use rewards intermittently and do not reward for every correct response.
- Too much handling may lead to boredom, too little will produce a scared or reluctant horse.
- The trainer needs to be 'smarter' than the horse. Problems mostly arise from the trainer and not the animal.
- Break up the tasks to be learned into simple, basic steps. Work from the known to the unknown.
- Always be consistent.
- Work within the animal's repertoire of ability. A horse will not learn anything it is incapable of doing.

Temperament
This is usually defined as the consistency in daily behaviour of any one animal. It is a complex of genetics and environment and has important implications in performance, general management and accidents to people. It is highlighted in the energetic contrast between a nervous thoroughbred and a quiet resting quarterhorse. Extensive work[4] attempting to relate temperament with performance in thoroughbreds and their offspring, has found no relationship.

From the practical point of view, the best approach is to assume that the main components of temperament in a horse are environmental so that how the animal develops is governed mainly by how it is handled and trained throughout its life. If the principles listed earlier are used, and a horse still has failed to respond then some genetic problem may be suspected, or some failure in the interaction between trainer and horse.

Early handling

A suggested training programme for a thoroughbred has been set out[15], starting from early foalhood, based on other research by Waring[22-24]. The main points from it are these:

- The foal is handled within 24 hours of birth to socialise it to humans.
- It is taken away from the mare for a few hours, spoken to gently while encircling it with the arms around the neck. It is only released when it has relaxed and is comfortable with people.
- At 2–3 days old, a soft leather halter is fitted and the foal is led when the mare is led. The foal is never driven.
- The foal is brushed and handled frequently and introduced to noises in varying environments.
- From 3 months old, the foal is haltered at regular intervals and led around vehicles, along farm tracks and so on, to get it used to different situations.
- Later training is done on a long rein, i.e. lunging to build up muscles through exercise. Here the horse learns the discipline of the rein and the role of voice command.
- Weight bearing should be delayed until bone and muscle development can take the strain, usually after 2 years of age.
- Further strange environmental noises can be introduced during this period (by tape recordings) to mimic real situations.
- Great care should be taken in correct mouthing of the horse using the correct bit and preventing soreness.

It must be stressed that this programme is one of gentling the horse from the day of birth. The approach is now to 'gentle' a horse rather than 'break' it.

Gentling the adult horse

Various horse handlers have learned by experience how the traditional 'breaking' of a horse can be replaced by the more effective gentling process. Examples are programmes of Mr Ray Hunt in U.S.A. who regularly demonstrates a 20-minute technique for gentling a horse. The so-called Jeffrey method of horse breaking works on similar principles and is described in reference 12.

The principle behind this is first to work within a comfortable space for the horse and then slowly to get the horse to accept your presence without fear or anxiety. Following a set sequence, the horse is roped around the neck, taught to lead, accept the saddle and be ridden. At no stage is there any need for pain, suffering or stress in the animal. Discipline is used but it is restricted to voice or minimal physical restraint of the animal.

Behavioural problems

Stalls

No other domestic animal has suffered such reductions in natural space as the horse. From being adapted to live in large areas of open range it is often confined to a stall of a few square metres. Little wonder they express their nervous energy in a variety of ways which in anthropomorphic terms are called 'vices'. These behavioural traits become self-rewarding and are hard to cure. The answer is to prevent them developing.

One researcher[19] considers that a stall subjects a horse to many insults such as draught, dust, unnatural lighting, temperature extremes and boredom. Other literature considers that vices or undesirable habits in stalls arise in bored horses with an anxious temperament. The problems are usually made worse by giving stalled horses high-energy diets without adequate roughage.

Some typical problems are:

- Stall weaving. The horse stands and shifts its weight from side to side in a stereotypic action.
- Stall walking. This is stereotypic walking about the stall also shown by caged animals in zoos.
- Stall digging, where the horse digs away at the floor with its front feet, usually in a corner.
- Bed eating – eating the soiled bedding (straw or shavings) in the stall.
- Wood chewing, where wooden parts of the stall are bitten into and chewed. This can lead to impaction colics, damage to the stall and arsenic poisoning. Lining the edges of troughs and doors with metal or creosoting the timber will reduce the problem. However, some horses develop a taste for creosote. Neck straps may or may not work.
- Crib biting or cribbing, where the horse fastens its teeth on to the wood of a trough or hay rack, arches its back bringing its head and neck to the vertical position while sucking air into its stomach. It may also be called wind-sucking. Surgical intervention is used to correct persisent animals but this may not be effective.

- Kicking the stall wall or door. A 30 m chain fastened to the back leg by a strap is used to prevent it.

Changing the animals' environment such as turning them out into a pasture relieves these problems but does not cure them. They usually revert when they are stalled. Kickers and cribbers easily pass on their stall responses to other horses nearby and occasionally when one vice is cured, another is learned. The solution is to reduce boredom and give the animal something to do. Examples are to feed high-fibre roughages to prolong eating, and feed this late in the evening to occupy the animals overnight. Provide distractions such as a beach ball hung from the ceiling of the stall. Give more exercise and allow them to see what is going on.

Two systems that have been used for exercising horses are shown in fig. 34.

Outdoors

There are some habits that initially start because of fear or are an exaggerated fright response[6]. Examples are:

- Sweating, shivering or rolling the eyes.
- Bolting, a flight response.
- Rearing and bucking, responses to dislodge a predator and resist being ridden.
- Biting and kicking.

Other so-called vices include stumbling, toe-dragging and resistance to having harness fitted. There are others associated with work such as failing to pull, failing to stop when backing and so on.

Management and handling advice

Suggestions based on behaviour studies[25,26] and from general experience for general management are these:

- Keep horses in groups of the same age.
- Have more feeders than the number of animals and do not place them in corners so that fleeing horses can become trapped.
- Allow strange horses to make their first contact across a fence before mixing them.
- Remember that horses which are apparently friendly in stalls may not be at pasture. The opposite is also possible.
- Mares should always be able to see their foals if separated for handling.

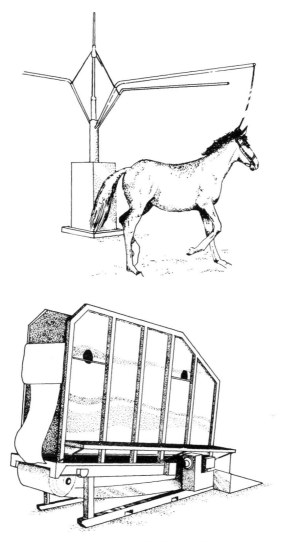

Figure 34 Two exercise methods used for horses. Rotating arm and treadmill
(after Fraser (1980) – see Further reading, p. 293)

- Fences and dangerous projections should be made clearly visible
  to horses.
- Door latches, locks and bolts should be protected as horses play
  with them and can learn to open them very easily.
- Keep horses occupied. Boredom causes vices.
- Ensure that catching and handling are pleasurable routines for a
  horse.
- Horses are large animals and cause many accidents to handlers.
  They should be treated with respect and care. Particular care is
  needed when horses are handled by children.

- Take special care when handling horses in narrow and confined spaces such as leading them into boxes, vehicles and aeroplanes. Give the animal time to see where it is going.
- Talk to horses in quiet, reassuring voice tones.
- Wear correct (protective) footwear to avoid foot injuries.
- Use the correct-sized gear (e.g. stirrups) and keep all gear in good order.
- Approach a strange horse towards its shoulder as it can see you more clearly than from the front. When familiar with the horse approach it from the front but hold out the hand to let it sniff you while talking to it.
- Horses do not like contact along the head, under the belly or the flank. Touch the legs and feet last. Run a hand along the horses' back or belly on the way down to the legs or feet.
- Recognise the signs of threat in the horse as warnings of their reactions. As man is a social partner with the horse, he too should give signals of his intentions.

Welfare

With growing interest in the horse for racing, recreation and sport, the welfare of the animal must be given high priority. The main areas of concern for people to meet the horse's needs are these:

- Feeding must meet the animal's needs. It is of special concern in the urban horse which is often kept in over-grazed, disease-prone paddocks. Adequate clean pasture is important with balanced supplementation when required.
- Stabling must provide shelter and shade and must also provide safe handling for the staff.
- Riding and working horses before they are physically mature can cause stress and injury to the animal.
- Fences should be made so that horses can see them (especially electric fences).
- Jumping horses at hunts and steeple chasing requires adequate preparation and training of the animal and experienced riders.
- Poorly designed race tracks that cause excess physical strain to horses need upgrading and adequate facilities to provide health care after injuries are essential.
- The use of whips, spurs and severely restrictive bits and harness needs vigilance by all horse owners to avoid the excess strain that their use can produce. Horses can be reprimanded without recourse to cruelty.

- Horses need adequate exercise and should not be kept in isolation.
- The feet of horses need special attention and their health should be regularly checked by experienced people.

Problems can best be prevented by people responsible for horses becoming more acutely aware of the animal's basic behavioural needs. Codes of practice have been published in several countries and an example of this type of information is presented below:

Code of practice for tethering horses and ponies
(a) Well-grassed, especially if grass is going to be sole source of food and free from poisonous plants, shrubs and trees, e.g. ragwort, laurel and yew.
(b) Reasonably flat.
(c) An area of shade available. Essential during summer.
(d) The area of tether must not cross a public footpath or be close to a public highway, especially where there is fast-moving traffic.
(e) The ground must not be waterlogged.
(f) There must be no large objects within the area of tether, e.g. tree stumps or bushes around which the tether can get tangled.
(g) Size of the site must depend on
   (i) the amount of grass available
   (ii) the frequency with which the horse will be moved.
   Minimum size should be a 7 m radius.

Type of tether
(a) A leather head collar or wide leather neck band fitted with a swivel.
(b) Must be firmly staked.
(c) Metal chain of appropriate weight for the animal to be tethered.
(d) Attached at stake with 360° swivel fitting at ground level, allowing pony to cover a complete circle without tangling the tether.

Age and condition of horse
(a) Horses and ponies less than two years old must not normally be tethered.
(b) Mares in season must not be tethered near stallions and vice versa.

Clean water
Must be available at all times in trough or container that is securely placed and not easily upset by the animal.

Food

Supplements must be given in all cases where grazing is not adequate.

Frequency of inspection

Horse and tether must be inspected at least twice in 24 hours.

Changing the site

Must depend on type of ground, amount of grazing, size of horse or pony, etc.

Weather

Horses must not be tethered in extreme weather conditions, such as a heatwave, frost, snow, driving wind and rain.

References

1. Arthur, G.H. (1967) 'Parturient behaviour in the mare.' *Proc. Soc. vet. Ethol.* **1**, 8–10.
2. Berger, J. (1977) 'Organizational systems and dominance in feral horses in the Grand Canyon.' *Behav. Ecol. Sociobiol.* **2**, 131–46.
3. Dixon, J.C. (1965) 'Pattern discrimination, learning set, and memory in a pony.' *Rept Dept Psychol. West. Res. Univ.* pp. 359–63.
4. Estes, B.W. (1952) 'A study of the relationship between temperament of Thoroughbred broodmares and performance of the offspring.' *J. Genet. Psychol.* **81**, 273–88.
5. Feist, J.D. and McCullough D.R. (1976) 'Behaviour patterns and communication in feral horses.' *Z. Tierpsychol.* **41**, 337–71.
6. Fraser, J.A. (1969) 'Some observations on the behaviour of a horse in pain.' *Brit. vet. J.* **125**, 150–51.
7. Ginther, O.J. (1979) *Reproductive Biology of the Mare. Basic and Applied Aspects.* Michigan: McNaughton and Gunn, Inc. 413pp.
8. Grzimek, B. (1945) 'Ein Fohlen, das kein Pferd kannte.' *Z. Tierpsychol.* **6**, 391–405.
9. Heird, J. (1981) 'Horse behaviour and training.' *Pony of the Americas* **26**(8), 32–3.
10. Hinton, M. (1978) 'On the watering of horses: a review.' *Equine Vet. J.* **10**(1), 27–31.
11. Houpt, K.A. (1980) 'The characteristic of equine sleep.' *Equine Pract.* **2**(4), 8–17.
12. Kirk, D. (1980) *Horse Breaking made easy. Based on the Jeffery Method.* Queensland: D.H. Kirk, 28 pp.
13. McCall, C.A., Potter, G.D., Friend, T.H. and Ingram R.S. (1981) 'Learning abilities in yearling horses using the Hebb-Williams closed field maze.' *J. Anim. Sci.* **53**(4), 928–33.
14. Montgomery, C.G. (1971) 'Some aspects of the sociality of the domestic horse.' *Trans Kansas Acad. Sci.* **60**, 419–24.
15. Montgomery, E.S. (1973) *The Thoroughbred.* New York: Arco.

16. Potter, G.D. and Yeates, B.F. (1977) 'Using principles of animal behaviour in horse management.' *Proc. Horse Prod.* Short Course Texas A & M Univ., pp. 204–14.
17. Price, E. and Mader, D. (1979) 'Horse learning ability tested.' *California Agric.* 33(4), 16.
18. Rossdale, P.D. (1968) 'Perinatal behavior in the thoroughbred horse.' In *Abnormal Behavior in Animals*, ed. M.W. Fox, Philadelphia: W.B. Saunders, pp. 227–37.
19. Rossdale, P.D. (1975) *The Horse from Conception to Maturity.* London: J.A. Allen, 224pp.
20. Schafer, M. (1975) *The Language of the Horse. Habits and Forms of Expression.* London: Kaye & Ward, 186pp.
21. Tischner, M., Kosiniak, K. and Bielanski, W. (1974) 'Analysis of the pattern of ejaculation in stallions.' *J. Reprod. Fert.* **41**, 329–35.
22. Waring, G.H. (1970a) 'Perinatal behaviour of foals (*Equus caballus*).' *50th Ann. Mtg Amer. Soc. Mammalogists*, 18 June, College Stn, Texas, 6pp.
23. Waring, G.H. (1970b) 'Primary socialization of the foal (*Equus caballus*).' *Amer. Zool.* **10**(3), 25 (abstr.).
24. Waring, G.H. (1971) 'Sounds of the horse (*Equus caballus*).' *22nd Ann. Amer. Inst. Biol. Sciences Mtgs*, 2 Sep., Colorado State Univ., Ft Collins, 3pp.
25. Yeates, B.F. (1974a) 'Applying principles of psychology to horse training.' *Proc. Horse Prod. Short Course*, Texas A & M Univ., pp. 77–93.
26. Yeates, B.F. (1974b) 'Recognition and use of social order in horse management.' *Proc. Horse Prod. Short Course*, Texas A & M Univ., pp. 71–6.

Further reading

Houpt, K.A. (1977) 'Horse behavior: its relevancy to the equine practitioner.' *J. Equine Med. Surg.* 1, 87–94.

Miller, R.W. (1975) 'Western horse behavior and training.' New York: Dolphin Books, Doubleday Co. Inc., 305pp.

Tyler, S.J. (1972) 'The behaviour and social organization of the New Forest ponies.' *Anim. Behav. Monogr.* 5, 87–196.

Waring, G.H. (1982) *Horse Behaviour: The behavioral traits and adaptations of domestic and wild horses, including ponies*, New Jersey: Noyes Publ, pp. 292.

Waring, G.H., Wierzbowski, S. and Hafez, E.S.E. (1975) 'The behaviour of horses.' In *The Behaviour of Domestic Animals*, 3rd edn, ed. E.S.E. Hafez, London: Bailliere Tindall, pp. 330–69.

7 Pig

The pig is farmed in a wide range of conditions around the world with the exception of Moslem countries. It was the first farm animal, apart from the hen, to be subjected to very intensive housing and management. As such, it has become the prototype for increased intensification in other farm livestock and is the centre for a great deal of welfare interest.

Behavioural definition

The pig prefers to live in families or small groups. Old males can be solitary. Social interaction including aggression is important when strange pigs meet. The sow, a nest-builder, is strongly attracted to a home and is clean.

Pigs prefer scrub or light forest to open range and have streamlined bodies for pushing through bushes. They rely on scent and hearing more than vision, and have a wide range of vocal communication. They have high ability to locate odour sources.

Pigs forage and root for food and eat a wide range of vegetable and animal products (including carrion). In temperate climates pigs have diurnal habits but in hot climates they tend to be nocturnal.

Juvenile males are tolerated for quite long periods within the family group. In natural conditions males fight only during the mating season. Before mating sows seek out the boar and the initial contact is snout-to-snout. The courting procedure is important in achieving effective reproduction. Mating is prolonged hence a rigid standing sow is essential.

Young are born in a nest and quickly set up teat preferences and show some rival social interaction. There is close body contact during suckling.

The pig is very sensitive to extremes of climate. It has no sweat glands, no thick hair cover and relies on a fat layer for insulation. Thus pigs huddle to keep warm and wallow to keep cool. Pigs are very curious animals and very adaptable. As a result they learn quickly.

Birth

Sow behaviour before birth

Observations[15] show that gestation length in the modern pig kept intensively varies from 108–22 days (mean 115 days). Sows carrying litters above eleven have a 5-day shorter gestation than those carrying average litters of nine to ten piglets. Small litters are carried longer than large litters.

Sows should be transferred to their farrowing quarters about a week before giving birth to allow them to settle into the new environment. Unless this happens, sows (especially inexperienced gilts) are prone to stress at birth and piglet mortality is higher. Heavily pregnant sows spend most of their time resting, sleeping and feeding. As gestation length varies widely, recognising signs of farrowing is important.

Signs of birth

Great variation is seen between sows in their behaviour before farrowing. This depends mainly on age, previous experience, breed or strain and the husbandry system. Some general signs are:

- Vulva swells and becomes more red, especially obvious in white-skinned pigs about 4 days before birth (range 1–7 days).
- Udder swells, becomes firm and colostrum can be obtained by gentle massage up to 24 hours before farrowing. Rubbing the front teats usually stimulates the sow to lie down.
- Increased restlessness. The sow gets up and lies down or changes side much more, twitches the tail, chews the pen railings. She also urinates, defecates and drinks more.
- The sow chews up bedding (if available) and makes a nest. She paws the ground, especially where no bedding is provided. This is a key sign.
- Discharge of blood, birth fluid from the sow and green-brown faecal meconium pellets originating from the piglets.
- Increased respiration from about 54 breaths/min 24–12 hours before birth, 90/min 12–4 hours before birth reducing to 25/min 24 hours after farrowing. This together with intermittent low grunting and jaw champing is common.
- Sow rectal temperatures rise from about 39°C to 39.5°C 4 hours prior to farrowing. They then stay elevated (around 40°C) for up to 24 hours after farrowing.

Farrowing pens and behaviour

There have been major changes in the last 20 years in housing and fittings for farrowing sows. Farrowing pens originally about 3.5–4 m² were reduced in area to become farrowing crates (2.4 m long and 700 mm wide) as weaning age was reduced. Floors have moved from partly to fully-slatted floors and then to wire and plastic slats. This evolution will continue with an increasing recognition of welfare requirements (see reference 49).

In farrowing pens or stalls, the sow is either restrained tightly or tethered so that she cannot turn round or move to and fro more than a few steps. No bedding is provided for nest building to reduce the task of cleaning.

In these conditions, the sow shows more heightened activity a few hours before farrowing which can lead to lacerations, bruising, abrasions and exhaustion. Slow parturition (uterine inertia) and lactation problems may result. Research[25] showed that sows in farrowing stalls sat up or stood up more and for long periods compared to sows in pens. The presence of straw did not affect the times spent sitting or standing.

It is considered that farrowing crates may save piglets from being laid on (overlaying) but by frustrating and stressing the sow and disallowing her maternal responses, overall productivity may not show an improvement. More research is needed, especially on the types of flooring and size of crates. Floors for farrowing are described in reference 49.

Sow behaviour during birth

During birth, the sow lies on one side, and in the intervals between births she may change sides, stand up or sit up in the dog-sitting posture. It is during these movements that risks of crushing the piglets are high.

Studies[35] show that the average time taken to produce a whole litter (averaging 11.3 piglets) is 2 h 53 min, or 15.3 min/piglet. The range in birth times/litter is from 25 min (3.6 min/piglet) to 8 h 55 min (44.6 min/piglet).

Normally, 55–75% of piglets are born head first and 25–45% back legs first.

Abdominal straining is more commonly seen before the birth of the first pig and less common with the remainder. As the sow strains, her tail is often pulled back away from the vulva, and delivery of a piglet is often accompanied by vigorous tail swishing and expulsion of gas from the rectum. Paddling with the legs while lying is common.

After all the piglets are born, the sow stands up and often urinates. Usually the foetal membranes start to be expelled during the birth phase and they may appear in two or three lumps. Most of the after-birth is shed after the last piglet is born. Four hours is normally needed to expel the afterbirth but this ranges from 21 minutes to 12 or 13 hours.

Overlaying by the sow

This is a serious problem and accounts for about 20% of all dead piglets. The greatest risk to piglets is during the first week of life and especially during the first few hours of birth when a restless sow gets up and lies down again.

The factors leading to death of piglets through overlaying are complex, inter-related and involve both the sow and the piglets.

The sow's previous experience, age, breed or strain and general health (which includes such things as lameness), leg and joint problems, obesity, skin parasites, mastitis or udder problems, all affect piglet overlaying.

The piglets may be dull, weak, inactive, unco-ordinated or suffer from coma. These all arise from chilling and starvation. Very low or very high birth weights can increase the chances of mortality. The runts of litters and any with splay-legs have a high risk of being crushed.

Piglet creeps

Attracting piglets away from the sow after suckling is the main method farmers have of reducing overlaying losses. Piglets are first attracted by light so dull-emitter heat lamps need a pilot light for the first 24 hours after birth to attract the piglets. Thereafter the heat alone will attract the piglets. Feed pellets are then placed near the heat lamp to start the piglets eating. Sows may often lie with their teats towards the heat lamp. Pressure on space may mean that creep areas are too narrow. The heated area then extends across into the sow's lying area, so that piglets lie near the sow and are crushed.

Savaging of piglets by the sow

Savaging is more common in gilts with their first litters than in older sows. A sow may snap at and injure an odd piglet, but seldom does the sow savage the entire litter. The first piglet born is most likely to be savaged as it seems that the sow is frightened by its movement and high-pitched squeak associated with birth pain. After the sow has sniffed the piglet, all is well.

Most sows will accept human presence at birth but some savaging may be the result of fear caused by stockman disturbance or other unfamiliar noises. Most breeders cull sows which savage litters and there is a suggestion that certain strains of pigs are worse than others.

Piglet behaviour after birth

Studies[35] showed that over 70% of piglets were born with their umbilical cords still intact and attached to the foetal membranes inside the sow's genital tract. The cord can stretch considerably before breaking. It took an average 2 to 6 minutes (range from 1 to 30 minutes) for piglets to free themselves from their cords which become shrivelled within 4–5 hours of birth. The sow rarely chewed the end of a piglet's cord.

After a brief period (5–10 seconds) of not breathing the piglet gives five or six gasps and a cough. Then there is usually about 20 seconds of rapid shallow panting followed by regular rhythmic breathing. Most piglets attempt to stand within 1 minute after birth and within 2 minutes can stand freely and start searching for teats or protuberances that feel like teats[44].

Interval from birth to first intake of milk ranged widely from 3 to 153 minutes (average 10–15 minutes) per piglet. Piglets clearly vary in their ability to find the teats although some are restricted by the trailing cord. Some suck the end of the sow's vulva which is like a teat. Nuzzling seems to be important in teat seeking.

Piglets show a very clear preference for the front teats of the sow and the first-born usually establish the best position in this 'teat order'. There is intense competition and fighting among the piglets until the teat order is established and stabilised.

Rearing

Sow and litter

The interactions between the sow and her piglets are complex. In a simplified form developments occur like this. The early-born piglets are attracted by the sow's voice to the front teats[34]. Early-born piglets, those of good birth weight (not necessarily the same) and those of greater vigour claim front teats and fight to defend them. Studies[28] recorded an average of eight fights during the second hour after birth reducing to two fights/hour by 8 hours after birth.

Even piglets reared on artificial sows with rubber teats establish a teat order[34] and soon learn to recognise their teat when it is moved.

The front teats have most milk and when massaged by the piglets or the stockman, milk let-down can be stimulated. Massaging the back teats is much less effective.

The front teats are longer with more space between them, and they have a greater clearance above floor level than rear teats so the piglets can grip them more easily. Clearance declines up to teat six. As the sow ages, the udder becomes more pendulous and her ability to expose the bottom teats is reduced[15]. The risks of being kicked by the hind feet of the sow are also greater at the rear.

The front-sucking piglets grow faster and consequently maintain their social ranking in the litter. Once a piglet has established its teat, it retains it. As let-down is very rapid (about 20 seconds) and occurs about once/hour there is no opportunity for piglets to share teats but they may suckle more than one teat. Normally, stockmen select females for breeding with a minimum of twelve functional teats (i.e. not blind or inverted nipples) and preferably fourteen to sixteen. In older sows, damage and disease may reduce the number of functional teats.

Behaviour studies[37] of piglets during the first 5 weeks after birth showed that whether they rested under the heat lamp loosely or tightly packed depended on the ambient temperature and there was little consistency in the position or orientation of the piglets. They lay down either by collapsing or by assuming the dog-squat posture followed by lowering head and shoulders. This contrasts with older pigs that lie down fore-end first. Piglets sleep for about 26 minutes in each hour.

The sow and litter accommodation should be designed to provide:

- A safe zone for piglets to rest and feed away from the sow.
- An interaction zone where sow and litter feed and socially interact.
- Space to allow the sow to feed, drink, rest, excrete. This area often overlaps with the interaction zone.

These requirements from the production and welfare viewpoint are discussed in detail in reference 49.

Vocalisation by the sow is important in piglet suckling behaviour. Studies[53] showed that piglet suckling behaviour normally progressed through the following stages (1) jostling for position, with squealing, (2) nosing the udder, (3) slow sucking, (4) rapid sucking, (5) final slow sucking and nosing (See fig. 35). The sow's grunting increased greatly up to the start of stage three which was a clear indication of impending milk let-down that started 25–35 seconds later. Recording

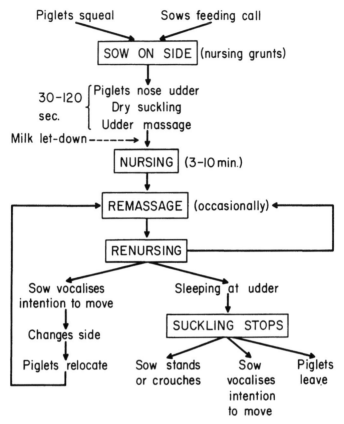

Figure 35 Sequence of responses during bouts of suckling in the pig (after Sinclair (1970), Purdue University)

these feeding sounds and playing them back to sows and litters can increase suckling rate, growth and weaning weights.

Isolated, lost or handled piglets vocalise loudly and this can alert and stress the sow. These piglets calls vary from closed-mouth grunts, open-mouth grunts to squeals. Stockmen should be in a safe position protected from the sow before picking up piglets.

Fostering

Where sows are farrowed separately but in batches, litter sizes can be adjusted by taking piglets from large litters and giving them to sows with adequate milk and small litters. The sows can react aggressively to fostered piglets so care is needed in the process. Best results are obtained if piglets from both sows are the same age and under 1 week old before the teat order is established. First remove the piglets from the sow to be given the extra piglets until they are all hungry and she is anxious to have her litter back. Rub the rear ends of the removed piglets over the strange piglet's anogenital area as the sow usually sniffs this area. The afterbirth may be used effectively. Watch the new litter during suckling to ensure that the sow accepts all the new members.

Fostering will disrupt the whole nursing process because of renewed competition for teats, and growth performance of the whole litter can be affected[32].

The outcome of sow and litter suckling problems is usually reflected in wide variation in the weaning weights of the piglets. This can affect their subsequent growth, marketing and carcass composition.

Cross-suckling

When sows farrow in batches in the open, litters may mix freely from birth. There are few acceptance problems and care is only needed to see that the piglets are distributed evenly among the sows or there will be wide variation in weaning weights.

Germ-free piglets

Germ-free (gnotobiotic) pigs are born by hysterectomy and live with one or two litter mates in isolation chambers away from any contact with their dam or other pigs. Mortality can be higher than expected from these piglets and behaviour studies[43] showed that compared to pigs born and reared normally, germ-free piglets were initially more restless and squeaked and squealed much more. It was concluded that the germ-free pigs' early environment was more stressful because of more handling, temperature changes, delays in initial feeding and absence of maternal bonding and reassurance.

The runt

In studies[40] of the runt problem, the first difficulty is to define in precise terms what a 'runt' is. A runt is not necessarily a small piglet as these can be the result of defective nutrition. These 'nutritional runts' are commonly seen in large litters where pigs are not weaned until 5 or 8 weeks. They can be early weaned and artificially reared or fostered on to other sows suckling litters as soon as they are identified. A number of them can be collected up and put on to a newly weaned sow that still has some milk and will take them.

Weaning

Pigs were traditionally weaned at 8 weeks of age, but in more intensive systems, 4-week weaning is usual. Some authorities consider weaning at 5 weeks to be best, while others prefer systems based on 3-week weaning. The effect of weaning at 24 hours when the piglets have only had colostrum from their dams is being studied.

Weaning has become an abrupt process as gradual removal from the sow is considered by farmers to be too costly in time and space. Whatever age piglets are weaned or whatever system is used they are subject to stress for the following reasons:

- The mother, their source of feed, warmth, security and comfort, is removed.
- They lose their normal clues for feeding given by the sow's grunts and squeals.
- They may be mixed with strange pigs in another pen.
- Competititon for feed, water, sleeping and dunging area may be increased.
- They will be subject to new bacterial challenges and disease.
- They may be subject to rough handling, transport, markets, veterinary inspection and treatment.
- They will experience climatic changes, cold and draughts being especially dangerous.

Piglets can generally cope with one or two of these stresses but compounding as in some present-day weaning procedures leads to poor growth for up to 14 days after weaning.

Pig farmers try to lessen this by ensuring that the piglet is consuming adequate quantities of a suitable diet (similar in composition to sow's milk) before it is weaned. Newly weaned piglets also take a few days to regulate their feed intake. This is seen in their defecating frequency which increases up to the 4th day after weaning and then levels out.

These problems of feed intake and digestion can be partly overcome by getting the piglet to eat more before weaning. The flavour, physical form and composition of the diet are involved though most diets are now fed as pellets which are palatable and eaten without waste. The settling-down period of weaners appears to be about 3 days.

Most husbandry problems in cages appear to relate to temperature control. Pigs weaned from 16 to 25 days (5–7 kg) leave the sow's creep nest at around 28°C and enter the growing pen at 25°C. The temperature is then gradually reduced to about 19°C before moving to the finishing house at a lower temperature. Behaviour problems and stress arise if the temperatures are too low or erratic as pigs have to huddle and fight for positions to keep warm. Feeding space is also an area of potential behaviour problems.

Research[5] showed that supplying caged pigs with a box and solid floor area provided a desirable environment. They invariably lay on the solid floor up against a solid wall.

Piglets which spend a very high proportion of their time on their feet as they have trouble in lying down comfortably are frequently disturbed. The so-called 'I-can't-get-comfortable' syndrome will add fatigue to the other stresses of weaning.

Wire cages and flat decks

Along with early weaning, systems have developed where pigs are kept on wire floors in cages, and then may proceed to flat decks.

Pigs may be weaned into cages as early as 7 days old. Cages are about 1200 mm long, 600 mm wide and 390 mm high with about nine pigs/cage. The cages may be in tiers of three to four units high and the pigs are kept at 27°C in subdued light.

Flat decks are pens with totally perforated floors and pigs can be kept in these from 7 kg liveweight through the growing and fattening stage. The perforations in the floor mean that a dung channel is not needed.

In semi-tropical environments pigs may be reared in cages without controlled environments. Farmers who use these systems claim the major advantages are to increase productivity through intensification while providing the piglet with a healthy environment. Details of this system and its welfare implications are fully described in reference 49.

An ideal group of pigs is nine to twelve and the fewer to the pen the better. With fifteen to twenty pigs/pen, performance declines. The recommended floor-space allowance is 0.24 m² /25 kg of pig and adequate alarm and safety systems in the event of power failure are emphasised.

Conclusions drawn from the limited behaviour research are often conflicting. One study[51] with 3-week weaners in cages found a high incidence of massaging and sucking other piglets and inanimate objects as well as dog sitting. This led to conflict among the pigs. In contrast another study[55] found no such abnormal behaviour and aggression declined rapidly once the piglets settled down.

Many commercial producers are running successful operations and record improvements in performance, lowered death rates and a complete absence of vices found in conventional systems. It seems clear that piglets adapt fairly quickly to cage systems and have similar play patterns to those reared conventionally.

Reproduction

The male

Boars reach puberty about 6 months of age although some variation can be caused by different feeding and management systems. Generally boars are not used for service until they are 7−8 months old.

With modern production systems based on high growth rate and confinement rearing, there is concern on whether this could affect a young boar's sexual performance[16].

Boars while still suckling begin to show elements of sexual behaviour during periods of play. Large numbers in weaner pens may encourage this development and it is increased when different weaned litters are joined.

Research[31] has shown that boars reared from 3 weeks within physical and visual contact with other pigs show better courting and copulating performance than those reared in isolation. Also group-reared boars had superior courting behaviour to those reared in individual pens.

Stable homosexual relationships which last many months can develop between male pigs reared intensively, although both members will copulate with females too. High stocking density stimulates aggression as well as abnormal sexual behaviour among intensively reared boars.

Before mating a sow, the boar starts a courting ritual where he chases the sow around nuzzling at her head, flanks, shoulder and anogenital area. He pushes or leans on the sow on occasions to test if she stands still and will drink her urine. During this period the boar urinates frequently, grinds and chomps his teeth, salivating freely and froths at the mouth.

This courting behaviour has been shown[29] to have an important effect on improving the conception rates in sows mated.

In extensive units boars often have rings or twisted wire inserted in the rim around the nose to prevent them rooting up pasture. It should be remembered that this will also prevent the boar nuzzling the sow and hence affect the courting ritual and subsequent reproduction.

When he mounts, the boar rests his belly along the sow's back and grasps her with his forelegs. Inexperienced boars front-mount, side-mount and dis-mount frequently before intromission and full protrusion of the penis. Ejaculation occurs when the cork-screw penis locks into the sow's cervix. Ejaculation is a fairly long process (up to 25 minutes, average 7 minutes). Compared to some species the boar produces a large amount of semen (500 ml). During copulation he thrusts several times and has rest periods and he may paw the sow's back. The sow must stand still for many minutes. Shortly after the completion of ejaculation he will dismount and withdraw the penis into its sheath.

The importance of pheromones in the boar's saliva and from his sheath have been extensively researched as has the importance of his mating grunts. To assist receptivity the order of priority seems to be smell, sound, sight and contact with the boar.

In boar management there are a number of behaviour aspects that should be considered. These are:

- Treat boars as individuals. Handle carefully and de-tusk them every 6 months. Remove the front accessory claws to protect the sow during mating.
- Do not overwork the boar. A boar to sow ratio of 1:20 is most common.
- Use sparingly. Four services a week are adequate until he is a year old, then he should not serve more than 6 times a week. Studies show that average litter size increases by 0.1 piglets for every extra day's rest before mating.
- Young boars should be first mated to older sows at peak oestrus. They should be supervised to ensure they are not injured and serve correctly.
- After lay-off periods longer than a month, a boar may have lowered fertility and reduced libido and his interest may have to be stimulated by using a sow at peak oestrus that has already been served by another boar.
- Spreading some ejaculate from another boar along the sow's back helps to stimulate the boar.
- The boar should mate the sows in his own environment (pen or yard). Otherwise he will waste time in an elaborate ritual of urinating, rubbing scent from his body into the walls, mark his territory with his salivary foam and fight the sow to establish dominance.
- The floor of the service pen should provide a good foothold for the boar but not too rough to cause foot soreness.
- Boars should generally be kept within sight, sound and smell of other sows. However, this assumes the boar is dominant to others in the herd and as boars are solitary animals in the wild, some individuals may be stressed by more dominant animals (e.g. another boar) housed near.

- Boars should obtain regular exercise to maintain fitness. Many systems allow for the boar, housed separately, to walk daily to the sow's accommodation to help stimulate oestrus and identify sows on heat.
- Boars should be handled regularly by stockmen walking behind with a board for protection from the boar's teeth. Boars respond to quiet handling and talk from the stockmen.
- Too frequent use of a boar as a teaser may frustrate him too much and he will not serve when needed.

Boar behaviour problems

Mating problems of boars are usually caused by bad early experience. Some of these are:

- Serving into the rectum instead of the vagina. Young boars should be supervised and helped if necessary to achieve correct alignment with the sow and effect intromission.
- Extremes of heat may make the boar unwilling to work. Mating should be delayed until evening although in hot climates some boars that will only mate at night upset normal piggery management.
- Masturbation by coiling the penis up into the diverticulum of the prepuce. Careful checks should be made to ensure the boar's penis has actually entered the sow.
- Boars that persistently masturbate should be culled although the diverticulum can be surgically removed.
- Some boars behave normally up to the point of mounting and then squat down on the floor and ejaculate. Great care is needed to help these animals to mount and not fall off as they may develop this habit.
- Aggression. Boars are always potentially dangerous and need care in handling. Animals that are nervous and become aggressive should be culled. Boars may show aggression with strange handlers while being quiet with their regular handler.
- When strange boars meet they strut shoulder to shoulder, heads raised and hair bristling along their backs. Deep grunts, jaw chomping and mouth frothing continues. In a fight, boars face each other with their shoulders in opposition and apply sideways pressure. They circle around slashing at each other with teeth and tusks. They may charge with mouths wide open and bite. The loser turns and runs away squealing. Subsequently dominant boars

need only grunt to achieve submission. Newly mixed boars fight less if they are both put in a strange environment.

The female

Puberty in modern fast-growing strains of pigs kept intensively depends more on age than on attaining a certain weight which can be affected by feeding level. Most gilts come into oestrus between 170 and 220 days of age, usually after they have been moved from the finishing pens at around 90 kg liveweight.

Research[12] has shown breed, season of the year and social environment (e.g. crowding) to be important in affecting the onset of puberty. The mixing during transport of gilts stimulates the onset of first oestrus and this is often exploited by stockmen. It seems that the physical motion in transportation is not the factor which triggers puberty but the stimulation of reproductive hormones through the stress of mixing with strangers.

Contact with male pigs is also important in stimulating first oestrus. However, if exposure is too early and for too long the gilts may become accustomed to the boar's presence and not react. The male effect can be used to synchronise oestrus by exposing gilts to a boar at about 160–175 days of age. Oestrus will occur in 60–90% of them within 10 days of exposure. However, pre-conditioning of the gilt by isolation from males is necessary before this will happen. Rearing gilts with contemporary boars has no effect on the age at which they will reach puberty.

In general practice, gilts are mated at their third heat period when they weigh around 118 kg. However, mating at the second heat has some economic advantages.

Oestrus

Gilts and sows in oestrus may show any of the following signs:

- Swollen and reddened vulva about 2–6 days before oestrus.
- Mucous discharge from the vagina.
- Restlessness and a poor appetite.
- Female coming into oestrus may sniff the genital areas of their pen mates. They may ride others and stand to be mounted.
- When in oestrus most females show the 'stance reflex' when they arch their backs, stand rigidly to being pushed and allow the stockman to sit astride them.

- Prick-eared breeds carry their ears erect and held back.
- Females in oestrus make a characteristic grunt or 'roar' and seek out other pigs while looking for a boar.

Standing oestrus, when the female accepts copulation, lasts on average about 48 hours (range 38–60 hours). However some sows may remain on heat for up to 120 hours. First oestrus is usually shorter than later ones and sows have longer receptive periods than gilts. Length of oestrus varies between breeds, and is affected by seasons and management systems. Ovulation occurs during the second half of oestrus and so most managers serve females twice, at the start and end of standing heat. Sows not mated cycle on average 21±3 days.

Considerable study[47,48] has been done on the 'stand reaction' of the oestrous gilt and sow. The intensity of the reaction increases as oestrus proceeds and this can be increased especially by smell (boar pheromones), sight and physical contact as shown in table 3.

Table 3 Factors involved in the 'standing' response of a sow in oestrus[47]

Stimuli	Sows responding (%)
Presence of active boar	100
Hand pressure plus	
boar out of contact	97
boar out of sight	90
no boar, boar odour spray	81
" " male preputial secretion	80
" " tape of grunts	75
Gilts, hand pressure	
isolation reared plus sexual contact	47
normal reared " " "	46
isolation reared	42
Hand pressure – start of oestrus	40
24–36 hrs later	60

Broadcasting the tape-recorded courting grunts of the boar will greatly stimulate the female. Even a short sequence of sound is effective and the rhythm and frequency seem to be the important aspects of the sound.

The natural odour of boars and their effect has been widely researched and reviewed[7]. Synthetic 'boar odour' is now available commercially as an aerosol spray. It can also be produced by warming in an old pan some of the gelatinous exudate from the boar's prepuce.

Some sows, and often individual gilts show a preference for a particular boar. This can cause problems in planned mating and service crates may have to be used to forcibly restrain the sow for mating. The resultant litter size may be adversely affected by the stress generated in the female.

Gilts often only stand for the boar in the presence of another gilt or sow in the mating pen, or with a familiar pig in an adjoining pen. However, if these other pigs providing company are nearing oestrus they will distract the boar.

It is recommended practice to mate gilts with an old experienced boar, and use older sows to train young boars.

Effects of housing

The important contribution of male and female behaviour during pre-mating is not generally recognised. The female assumes the major role in searching for the male, converting social contact to sexual behaviour and releasing the mounting reaction. In contrast the male waits for the behaviour signals from the female[47].

Some housing systems arrange the boar pens between pens of six sows to achieve maximum physical presence while in other systems the boar is walked to the sow pens daily to test for oestrus[16]. These are criticised as being wasteful of labour. If the boar is put in with a group of females (perhaps closely confined) this too can waste time as the boar investigates the new environment and all the females at random within it. A boar may pursue females not in heat and ignore those willing to accept him. In intensive systems the responsibility for deciding which sows are on heat rests mainly with the stockman and his records. The most likely sows are brought to the boar in his own area and he confirms or rejects them.

With stalled sows the value of walking boars along the front to allow nose-to-nose contact can be questioned. It is time-consuming and inaccurate but allows the 'boar effect' to get to the sow. It is a natural approach. Walking the boar behind the stalled sows can be equally inaccurate as the boar is confused and may see the immobilised sows as an invitation and visual signal to mount. Sows should not be served in their stalls as injury can occur to both partners and the lack of pre-mating courtship may affect litter size.

Detecting sows in heat must remain the responsibility of the stockman and sows are then taken to the boar for checking.

The breeding female

Oestrus in the breeding female usually occurs within 7 days of traditional weaning of the litter, whether this is at 3 or 5 weeks weaning. Earlier weaning can allow more litters per year and greater productivity from each sow.

The gilt weaning her first litter usually has a longer return to service than older sows that have had many litters.

The re-breeding interval in sows can be shortened by improved feeding (flushing) to regain live weight and hormone treatment. It can be lengthened by stress usually caused by the fighting which occurs when weaned sows are re-grouped.

Induction of oestrus during lactation without resort to hormone treatment is often achieved by removing the piglets from their dams for 12 hours each day from about 21 days after farrowing. Also a number of sows and their litters may be grouped together from 15 to 25 days after farrowing in the presence of a boar with *ad libitum* feeding. Varying degrees of success are reported from these techniques but they are still being perfected. Some treatments may not induce oestrus during lactation but shorten the re-breeding interval after standard weaning.

Artificial insemination (AI)

This is now well-established in commercial pig production. As correctly diluted semen has a keeping time of about 3 days, it can be despatched by post for sow insemination by trained stockmen.

Boars can usually be quickly trained to mount a rigid dummy. Similar mounting behaviour is shown to the dummy as a female including nuzzling her flank and hind end. A sexually mature boar will mount the dummy and serve into an artificial vagina. Libido varies between boars and is affected by frequency of collections. Boars can be stimulated both by allowing false mounts, or by observing a collection from another boar. Patience by stockmen is important.

Some boars work best if their first sexual experience was on the dummy with an artificial vagina. Boars that have first mated sows may be more difficult to train. However, boars vary greatly and some seem to need a sow to mate before they will mount a dummy. Old experienced boars will usually mount the dummy without hesitation.

Using a live sow and an artificial vagina is not normal practice as the sow adds another complication to the routine.

For insemination a rubber catheter with an anticlockwise spiral on the end is inserted into the vagina and gently rotated until it locks into the cervix. Semen is run in by gravity through the catheter. This takes about 5—15 minutes. Success depends on correct identification of oestrus by the stockman. Some operators use the aerosol 'boar-smell' sprays, tape recording of a boar's courting grunts and the presence of a teaser boar near the sow. Other breeders inseminate the sow while an assistant sits astride her, although not all sows allow this.

The actual presence of a boar or male courtship is not essential in achieving good results from AI. Vasectomy is not necessary for a teaser boar as the time delay between mounting and copulation allows the female to be separated. However, this can result in a very bad-tempered boar and care is needed in handling.

The castrate (barrow)
Castration, known to reduce the aggressive behaviour of a boar, acts mainly by reducing the testosterone level which lowers the boar's sexual drive and his aggression. Castrates act like gilts and can be managed in a similar way. Any behavioural problems that arise in barrows are caused by their faster growth rate and competition for feed rather than their gender.

Production

Growth — fattening
Problems that can arise during this growth are inevitably expressed as poor performance, low growth and poor feed conversion efficiency[14,45]. The stockman has to decide which of the many components of management have been the cause of the problem, and it is within this often complex situation that behaviour can be implicated. Some of these factors are summarised in fig. 36.

Some of the major points are:

Feeds and feeding method
Pigs generally prefer wet rather than dry or dusty feeds. Dry feeds (meal or pellets) take more time to eat and hence provide greater 'occupational therapy' than the rapidly eaten wet feeds fed in troughs. The pig farmer decides on the physical form of the diet based on costs and feed conversion.

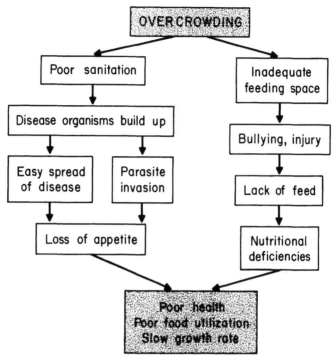

Figure 36 Possible pathways by which overcrowding affects pig production
(after Beck, P.C. (1962). *J. Agric. W. Australia* **3**, 721–5)

Other feeding decisions are whether to feed *ab libitum* or restricted, and whether pigs should be fed once or more times a day[8].

As a general rule, the aim should be to restrict the animal as little, not as much, as possible. This concept is generally misunderstood as most farmers think that restriction is giving as little as possible.

Pigs perform well on once-a-day feeding but this means that the number of daily inspections by the stockman will be reduced. Pigs should be used to stockmen walking through the growing house and not expect to be fed. In automated houses which are kept at low-light intensities, even putting the lights on for inspection disturbs the pigs and the noise level makes it an unattractive environment for the stockman unless ear protection is worn. These noise levels may also affect the pigs. Automation should in fact increase the stockman's opportunities to observe the pigs.

Stockmen should realise the importance of giving pigs adequate feeding opportunity. Pigs may be fed *ad libitum* yet if there is restricted access to the feed because of the design of the feeder, the pigs encouraged to eat through social facilitation will become frustrated and aggressive[26]. However, restricted feeding combined with inadequate space is even more likely to cause frustration than with *ad libitum* feeding.

Stocking density

Increasing stocking density (more pigs/pen) is a typical way to increase throughput and return on capital invested. However, the effects of too high a stocking density (overcrowding) are seen in bullying, vices (ear and tail biting), diseases, low growth and poor feed conversion which lowers throughput and profit. The main behaviour problems can be traced to alteration or upset of social order within the group and this has been studied for different ages and weights of pigs[18,19,41].

With increased stocking rates, the pigs spent more time feeding and less time sleeping and resting. Highly stocked pigs also spend more time standing and walking. Clearly, high densities greatly increase disturbance and tension and hence aggression. The space requirements for pigs as recommended by Bogner[6] together with suggestions for a controlled environment are presented in tables 4 and 5.

Table 4 Minimum space requirement of pigs[13]

Piglets kept in cages
   Minimum space (cm^2) = 87 x weight (kg) + 440
   Not more than twenty piglets should be kept in a cage unit
   The difference in weight should not be more than 20%

Weaned piglets ($<$ 30 kg live weight)
   Minimum space (cm^2) = 87 x weight (kg) + 440
   Length of trough (cm) (using restricted trough feeding) = 0.34 x weight + 12

Breeding and fattening pigs ($>$ 30 kg to 100 kg)
Group penning with separated dung and resting areas
   Minimum space (cm^2) = 30 x weight (kg) + 2000
   Length of trough (cm) (using restricted trough feeding or automatic
      feeding) = 0.15 x weight (kg) + 18.5

Breeding pigs
Single keeping (measures of the stand)
   Breadth   = 75% of height at withers
   Length   = 1.4 x length of body
   Partition = 15 cm distance partition to floor

Group penning with separated dung and resting area (100 kg)
   Minimum space (cm^2) = 90 x weight (kg) − 4000

      Compared to single keeping, for group penning without separated dung and
         resting area, the dimensions have to be enlarged by 20%

Table 5 Controlled environment for pigs[13]

	Live weight (kg)		Minimum temperature (°C)
Piglets kept in cages	2.5	− 5	26
	5	−10	24
	10	−20	22
	20 and more		19

Temperatures should not rise above 35°C
Air humidity should not rise above 80% or go below 50%.

		Without straw	With straw
Weaned piglets	10–20	22	18
	20–30	18	14

Air humidity should not sink below 50%

Breeding and fattening pigs		10	5

Temperature should not rise above 30°C. Air humidity should not rise above 80% and should not sink below 40%.

The climate requirements of pigs are reviewed in reference 13 and extensive details of environmental control are given in reference 10.

Practical observation shows that the most aggressive pigs are not necessarily the best feed converters as they use up a lot of energy on maintaining their social rank. More restful pigs are better performers.

Floors

From the pig's point of view natural floors (i.e. earth) provide much more than support, as they are readily manipulated to provide food and also novelty. The objectives of an ideal floor have been defined[3]. These are:

- To provide a non-slip, non-abrasive surface with no protruding edges.
- Not to harbour bacteria or parasites. Surfaces should be impervious and readily cleaned.
- Not cause stress or discomfort and should meet the needs of the animals living on them.
- Materials and structures should not deteriorate or deform during their planned life and not require maintenance.

- Perforated floors should not retain faeces or urine to the point where they need cleaning or scraping.
- Floors should meet the above requirements at the lowest acceptable cost.

The pig's cloven hoof was clearly adapted to walking on earth so it is not surprising that there are many documented reports of injuries to pigs' feet caused by slatted and perforated floors. Much of the damage can be attributed to inappropriate materials being used, along with poor design, manufacture, installation and maintenance.

Pigs prefer to walk on wide rather than narrow slats or perforated floors and recommendations for concrete slats in Europe and the United States come within the range of 75–125 mm slats and 18–25 mm gaps (see reference 49).

No real solutions have been offered for these problems and more research is needed on safe, non-slip surfaces that can be economically installed. Pigs that are frightened of slipping are under stress and this leads to poor performance and aggression.

Mixing pigs

As an alternative to flat decks newly weaned pigs are sometimes put into larger groups or weaner pools. From these groups pigs are drawn on the basis of size, age, sex and weight for the growing house. To reduce the stress associated with this the following suggestions are made:

- Mix three to four sows and their litters before weaning.
- Do not put newly mixed pigs in too big a pen as they will remain in their own groups. Most fights occur between groups and not between pigs from the same litter as they already have a stable social order.
- Place all the pigs to be mixed in a strange pen.
- Distract them by providing straw and paper sacks to chew or spray them with a strong-smelling fluid (waste oil or diesel).
- Provide adequate trough space and feeding and drinking opportunity.
- Mix them at dusk as identification of rival pigs is then more difficult.
- Keep them in a close weight range and remove any sick pigs immediately.

Producers often run batches of up to fifty pigs (25–30 kg) and then reduce them to groups of twelve to sixteen prior to slaughter.

This may not agree with recommended group sizes for flat decks of 9–11.

The fewer pigs in a single group, the tighter they can be packed. It is general experience that vices are more common in large groups in large houses where social orders are more complex and maintaining them requires more aggression and stress.

Diets should not be changed at the time of mixing and pigs should be fed *ad libitum* for 2–3 weeks after mixing to avoid digestive upsets. There should be adequate trough or feeder space. Most producers prefer *ad libitum* feeding up to 10 weeks of age.

Research[24] compared pigs kept in their litter groups with mixed litters on full and restricted feeding. Most fighting occurred among restricted-fed mixed groups. Encounters among litter mates did not lead to fighting whereas it did between mixed pigs. They suggest that restricted feeding should be delayed until new social orders are evolved in mixed groups to reduce stress.

Opinions about mixing sexes vary. Some producers advise separate housing of gilts from castrates as the latter have greater appetites, need restricted feeding and are more aggressive than gilts. However research[2] where entire male and female pigs were housed together from weaning to bacon weight showed no adverse effects on performance. During the last 4–5 weeks some mounting occurred but no pregnancies. Thus decisions on whether to mix sexes should be based on nutritional criteria and not behavioural ones as these effects are minimal.

Drinkers

Water is provided by water bowls, metal nipples that release water when they are pushed, or metal 'drinking straws' up which the pig sucks water. All are effective ways of supplying water but stockmen vary in their personal choices.

From the behavioural viewpoint, the watering system will also provide a plaything for the pigs hence it will be subjected to rubbing, nuzzling and chewing. Rigid construction is essential. In positioning waterers, points to remember are:

- Place near dunging area, near a drain or above a food trough because of spillage.
- Place nipples so that pigs have to reach up to them when drinking.
- Don't place in positions where pigs can bruise themselves.
- Place so that drinkers will not be damaged by rubbing. Siting in a corner will prevent this.
- Be consistent in the choice of drinkers used.

Dry sows

There are a number of different systems of keeping dry sows during pregnancy and the history of developments in sow housing is well described in reference 49. The main concern in the design of accommodation is to achieve economies in space, building material and labour as well as protect the animals and control their individual nutrition. The basic housing systems are:

- Sows locked into individual stalls by a restraining bar at the rear. They can move to and fro but not turn round. The environment is controlled.
- Sows tethered in stalls by a strap around the neck or a girdle around the chest. They cannot turn round and the environment is controlled.
- Groups of sows at pasture with communal shelters against cold, rain, snow and sun.
- At pasture in individual runs or tethered to individual kennels.
- Groups (six to ten) in a yard with a kennel-type shelter and individual feeders.

In groups, the main behavioural problem is the establishment of a social hierarchy and this is reviewed in reference 33, who also drew up an ethogram describing grouped sows' behaviour. These showed the social interactions between sows to be quite complex.

Considerable interest and welfare concern has been focussed on sows in stalls. In some systems, sows are even tethered during pregnancy and suckling. Stalls restrict the sow's movement (she cannot turn round) and she has little opportunity for exploratory behaviour while lying on a solid floor. Research[22] showed that adding straw increased the sow's apparent comfort and provided material to chew and so reduce boredom. Bar-biting is a regular pastime for stalled sows when disturbed before feeding or watering. It also occurs in a chronic form during the day in the absence of any special disturbance and before defecating and urinating. Provision of straw reduced this chronic bar-biting.

Behaviour in the 'Pig Park'

Three sows and one boar of commercial stock were run with minimal care in a 1.5 hectare 'pig-park' near Edinburgh for several years, while their 'normal' behaviour was recorded. Farrowing sows dig nests in the ground lining them with swamp vegetation or chewed-up bark or branches. As their oestrus is synchronised they have their piglets within a few days and soon take up a large communal nest.

The daily activities follow a set routine with all pigs moving along corridors from feeding to wallows and then to resting vantage points.

Based on normal activity shown by these animals, a new design for a commercial pig unit has been built and preliminary results indicate production is equal if not better than in current commercial units.

Using first principles to fit houses to pigs was the aim of the pig park study and the results appear to be very promising.

General

Sight

Pigs are basically forest dwellers and nocturnal feeders. They have an eye which is as highly developed for vision including colour as other higher mammals. A pig's sight can be greatly affected by its floppy ears which although they may add protection in scrub, clearly restrict vision.

Studies[21] have shown that sight is not essential in the formation and maintenance of social ranking in pigs as long as they can use their other senses. Sight is important in breeding as the boar in particular uses visual clues to assess the reproductive state of the female. However, pigs rarely explore their habitat from a distance, and do not rely on vision to any extent.

In housing systems where pigs have been allowed to record their preferences, young pigs (12 weeks old) recorded no discernible day–night preferences but kept the lights on for 72% of the total time. However, care is needed in interpreting such trials.

Smell

The pig has a very acute sense of smell and this is exploited by man for hunting truffles and buried drugs. The pig has a well-developed olfactory lobe in the brain. The design of the nose and strong neck muscles give the pig unique rooting abilities.

Feral pigs use scent as an important method of following trails, and their olfactory ability seems to override their dependence on sight. They can locate each other in the wild by smell when they are spaced out foraging. Smell and taste are thought to be important in recognition as at first meeting pigs sniff each other in the face, along the line of the jaw and around the eyes and ears. Smell is certainly important in piglets locating specific teats in the suckling order.

Hearing

Pigs have very good hearing and clearly recognise a wide range of different vocalisations. They are easily panicked by sudden loud noises such as aeroplanes or thunder.

Taste

Pigs have been shown to have certain taste preferences. Creep feeds for suckling piglets are sweetened with sugar to encourage intake and mature pigs show a preference for apple flavour.

Vocalisation

Pigs use a wide range of sounds to communicate with their companions, developed possibly because of the poorly lit scrub environment they evolved in.

Young piglets squeak, grunt, bark and squeal and older pigs use various grunts and squeals to indicate hunger, thirst, alarm, fear, terror, 'affection', calling the litter to feed, courtship and many others. Experienced pig handlers recognise these and suggest that pigs also respond to the human voice that can be used for soothing, calling, persuading and admonishing. By voice a bond can be established between animals and stockmen.

One researcher[9] suggests the use of 'uff, uff' uttered with the lips, the face level with and close to the pigs is a good noise to establish friendship. She suggests any noise if spoken softly to the pigs will aid friendship. To move pigs forward 'choo, choo' achieves results, and if they break back use the same sound but with a sharper tone while clapping hands sharply. These suggestions would justify further study.

Pig squeals can reach 112 decibels which can become a hazard to stockmen's hearing and possibly other pigs. The wearing of ear protection by the stockmen is advisable. In designing systems, noise levels of an 85 decibel limit should be aimed for. The feeding system can affect the amount of noise produced by growing pigs.

Comfort behaviour

Pigs have no sweat glands in their skin and use water and wallowing as a means of keeping cool in hot climates. However, pigs also wallow in cold climates so the practice may have a role in skin and hair care. If water is deep enough pigs will swim. The cooling effect of mud lasts much longer than wetting of the skin.

There is very little or no social grooming in pigs as seen in other farm species. This is surprising in an animal that enjoys scratching or rubbing so much. They scratch themselves by rubbing against walls, pens, trees, fence posts.

Pigs yawn like other farm animals, and this is often combined with stretching the back leg and closing the eyes. Pigs also shake themselves after a period of rest, especially if they have been lying in a group. They also 'tuck in' by wriggling to settle right down into their bedding material or sleeping position among the group. Whereas wild pigs show a clear seasonal pattern in their behaviour, this is not so obvious in intensively farmed pigs.

Where growing pigs have been allowed to select their own temperature by nose pressing, they tend to follow a diurnal pattern of temperatures. They prefer warm afternoon temperatures and quite cool conditions at night, when they huddle for warmth. Pigs will press for cold air at night if conditions are warm (see fig. 37).

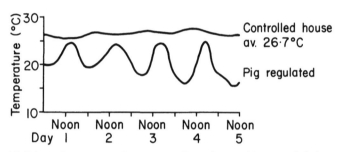

Figure 37 Diurnal patterns of response when pigs could control their own house temperature (after Morris and Curtis (1982) – see Further reading, p. 190)

Play

Pigs play frequently after they have left their mother's nest or lying area, especially if there is room to move and investigate the environment such as in outdoor situations. Play is often taken as a sign of general wellbeing of the piglets.

They play in groups (social play), on their own (solitary play) or with objects. In flat decks they often play in mid-morning. Social play usually consists of fighting with selected litter mates. Often one animal suddenly runs away and the rest race after it in pursuit. A piglet playing on its own often whirls around trying to scratch itself and may jump up and down on the spot. Pigs are great investigators and play by lifting and moving objects on the ground or throwing them in the air. Digging (nose rooting) is a favourite pastime and treasured finds are carried away, with other pigs chasing in pursuit.

Learning ability

Pigs have the ability to learn quickly, though there is a wide variation between individuals in their ability to learn and solve problems. The pig's learning ability has been shown to be moderately inherited so selection could bring about improvement of the trait. Pigs have been taught by various techniques (e.g. avoidance of electric shock, audio signals or feed rewards) to select temperatures and light regimes of their choice, or operate feeding and watering devices. In American studies, pigs when given the opportunity to select their own heat pattern chose a diurnal pattern so that heating costs were halved. Indeed trying to develop a pig-proof door and catch has caused considerable problem for many engineers.

Pigs work within a space dimension and quickly learn a new environment. Their ability to learn is part of their clearly recognised exploratory behaviour which is a species-specific trait. Depriving them of this causes problems and stress and it has been suggested that pens should be designed to provide more opportunity for exploration by pigs.

Dunging

Pigs are probably the cleanest and most orderly farm animals if their environment allows them to be so. They keep their sleeping area clean and dry and urinate and defecate in a specific area. This can be exploited in buildings designed to reduce labour and the energy cost of cleaning.

Studies[52] showed that piglets under 3 days old have no preferred site for dunging but after that they clearly avoid eliminating in the areas where they sleep. After this age they may learn from the habits of the sow as a 'dirty' sow will soon have a 'dirty' litter where they eliminate in the bedding.

Good dunging habits can be encouraged by management and certain correction of bad habits can be achieved. Here are some suggestions:

- When changing pigs from one pen to another, give them a run for a while before they enter the new pen to let them empty out.
- When moved, the dunging area of the new pen should be dampened and some dung from their old pen placed there.
- Pigs that are scouring on their bed should have the scour problem solved before trying to retrain them.
- If there are too few pigs in the pen, use temporary barriers so that

the sleeping area is well packed with pigs and they are forced to dung elsewhere. Put fresh bedding in their sleeping area.

- A false roof over the sleeping area will encourage them to sleep and rest there and they will be less keen to dung in their bed.
- Cleaning out at night is better than morning cleaning if the problem persists. After cleaning put some bedding or feed pellets where the soiling occurs.

The shape of the pen and the position of the feed and water source affects selection of an eliminative area. Dunging positions can be shifted by altering the food and watering points (see fig. 38).

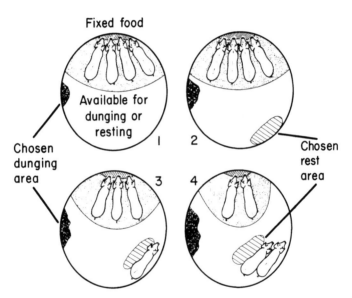

Figure 38 Behaviour of pigs in a round pen. 1, all pigs feeding, dung area selected; 2, rest area chosen distant to both; 3, first pig dunged and moved to rest area; 4, second pig likewise (after Petherick, J.C. (1982). Scottish Farm Buildings Investigation Unit Project)

Pigs appear to defecate and urinate regularly throughout the day and night in intensive housing, although there is a lengthy rest period between 7 p.m. until 6 a.m. the following morning. Pigs regularly align themselves with a wall when eliminating and do not defend a corner from their penmates.

Disturbed behaviour and stress
The abnormalities observed in pigs are summarised in Table 6 after reference 11.

Table 6 Abnormalities in the behaviour of pigs[11]

Type of abnormality	Suckling piglets	Early weaned piglets	Fattening pigs	Sows
Tail and ear biting	*	*	*	
Cannibalism		*	*	
Belly nibbling		*		
Tongue rolling		*	*	*
Rubbing nasal bone		*	*	*
Rubbing snout		*	*	*
Hyperactivity		*	*	*
Massage of anus			*	
Vacuum chewing				*
Biting bars of the pen				*

The factors that have been identified and studied as causing stress in pigs are many and varied, and many of them interact with each other. A list would include the following:

- Chilling
- Overheating
- Physical injury
- Poor sanitation
- Poor ventilation
- Overcrowding
- Bullying
- Dampness, draughts
- Genetic makeup
- Weaning
- Castration
- Lack of feed and water
- Nutritional deficiency
- Internal and external parasites
- Disease
- Loss of appetite
- Noise

These are all products of bad husbandry and poor stockmanship. Some producers say that these factors (or stressors) cannot be totally eliminated and it is not until they cause distress (in contrast to stress) that action must be taken.

Some of the common stress factors and their effects are discussed in the following sections.

Aggression – fighting
Aggression in pigs has been classified[17] as follows:

Acute: Fighting to establish a social order when strange pigs are mixed.

Chronic: Fighting to maintain an established order.
Abnormal: • tail biting
              • cannibalism
              • sudden savaging of a group member
              • sows attacking each other
              • sows savaging their young

Domestic pigs, in contrast to wild pigs, are kept in monocaste social groups of the same age, sex and size. This may in fact make it more stressful for pigs to form a social group when the combatants are physically equal.

Newly acquainted pigs seem to fight for 24–48 hours to establish a linear dominance hierarchy usually with the largest animal at the top and smallest at the bottom. Once the hierarchy is established, fighting is greatly reduced but not eliminated.

This fighting can be reduced by having adequate feeding and watering space as through social facilitation one pig can encourage the others to start eating and drinking. If there is no room then fighting starts. Hunger has been shown to encourage aggressive biting when unfamiliar pigs are mixed, with most biting occurring about 24 hours after withdrawal of the feed[36].

Giving straw to pigs fed *ad libitum* did not reduce aggression in this study but in another study[50] a handful of rye straw kept each pig occupied for 1½ hours each day and reduced stress reactions.

One of the most important practical problems is when individual pigs may have to be removed from a group for treatment and then returned. If they cannot be returned, keeping them separated becomes wasteful of space and labour.

In studies where pigs were removed from groups[20] it was found that top-ranking pigs could be safely returned to their group even after 25 days' isolation but low-ranking animals were severely attacked after only 3 days' absence. Clearly the stockman should have some idea of a pig's social rank before returning it if injury is to be avoided, but this will take time to determine. The social ranking that developed after 48 hours of mixing (when most fighting stopped) was not always completely stable. The physical size, previous experience, sex, genotype, environment and identity of other group members were all factors involved. It was not always the heaviest pig that became dominant, and careful observation and a lot of trial and error will be needed if stockmen want to modify social rankings in groups of pigs.

Moving and transport

Pigs are den-living, home-loving individuals and do not show a flocking response. They dislike being moved, especially from their darkened pen into the bright outside light, and easily panic. In the wild, frightened pigs scatter, crouch in the undergrowth or race back to their den. Domestic pigs still show this behaviour.

The deaths caused in pigs during transport can be of great economic significance and in some countries tranquillisers are administered before transport to slaughter. This is illegal in certain other countries.

Loading and unloading are the most stressful part of transport when pigs are pushed, dragged by the ears or tail, or goaded physically or electrically. Moving away from their own pen and mixing with other pigs in the truck causes trauma and fighting whenever the vehicle is stationary. Some suggestions to assist loading and transporting pigs are:

- Select the market pigs the night before transporting and hold them in an unfamiliar pen.
- Reduce feed but keep on full water.
- Move them in the early morning.
- Do not attempt to rush, abuse or punish pigs.
- They will go better up a ramp or along the level. Avoid having to load them down a ramp. Keep ramp sides fully covered.
- Spread some of their bedding up the ramp or use food pellets or cut slices of apple to encourage them to move freely.
- Use a hand-held board for coaxing. Do not beat them or drag by tails and ears. This panics the entire group and bruises the meat. Electric prodders should be used as a last resort if gentle coaxing fails.
- To move an awkward individual, put a plastic bucket over its nose and eyes and reverse it along.

Suggestions are drawn in fig. 39.

Transport deaths have become a recognised hazard in modern pig husbandry and as a result of studies[1,38] the following recommendations are given:

- Provide adequate ventilation in transporters.
- Avoid transport when temperatures exceed 22°C.
- Avoid long parking periods on hot sunny days.
- Avoid physical exertion and excitement.
- Do not feed during the 12 hours before transport.
- Avoid the use of stress-susceptible breeds and strains in commercial breeding programmes.

Figure 39 Pig handling techniques (after Ewbank (1968) – see ref. 6, page 279)

Slaughter

In some slaughtered pork pigs, the meat is pale, soft and exudes liquid (termed PSE meat or pale watery pork) and is of little commercial value. Although this happens to the meat after slaughter, it is caused by stressing the animal before slaughter[54]. The way to avoid it is to reduce stress between farm and slaughter.

A restraining conveyer for stunning would reduce the stress while nearing the slaughter point. This is better than having to chase and catch pigs for stunning or dragging them by tails and ears.

Research[42] has shown that when pigs from different producers were penned together, intense fighting took place in the first 30 minutes after mixing. After one hour, the pigs rested for the next hour and a half regardless of whether they were in single or mixed producer-lots. Holding pigs for too long in lairage will certainly increase stress as fighting recommences with increased hunger.

Breed

It appears that selection for leaner, fast-growing pigs has reduced the animal's ability to adapt to stress. Certain breeds and strains within breeds show stress in the form of sudden deaths, transport deaths and pale watery pork. Selection programmes are now underway to reduce this Porcine Stress Syndrome (PSS) by selecting replacements that remain relaxed during a brief period of anaesthesia using halothane. (See reference 49.)

Tail biting

Tail biting is a problem that has developed and increased with more intensive pig husbandry and starts usually 4–22 days after weaning. Ear biting may be associated with it. Many reasons have been put forward for the problem, yet little is known. They are:

- Behavioural – boredom, breakdown of dominance order, excess social contact.
- Nutritional – low fibre, low bulk, deficiencies of protein, iron, calcium, iodine, sodium chloride, unsaturated fatty acids are some points covered.
- Environment – poor ventilation leading to high humidity, high temperature, high ammonia and carbon dioxide levels. No bedding, inadequate trough space and watering points.
- Disease – skin mange, internal parasites, various infections.
- Teething problems.

Suggestions for preventing and reducing the problem include the following:

- Check ventilation to avoid build up of foul air and high humidity.
- Check that environment is not too cold (below 15°C) and does not rise above 28°C.
- Remove the aggressor(s) if it can be identified.
- Remove badly bitten pigs from the pen and isolate them.
- Daub the tails of all the pigs with Stockholm tar, creosote, disinfectant, lice or mange wash to act as a repellant.
- Provide chains, paper sacks, straw, balls, stone-filled cans for occupational therapy to reduce boredom.
- Check that feeding and drinking facilities are adequate and that bullying and fighting is reduced to a minimum.
- Reduce the stocking rate. Do not exceed 120 kg live weight/m² of floor space.
- Group pigs on the basis of similar size. Often a small pig in a group becomes an aggressor.
- Move problem pigs to another pen. This may either solve the problem or make it worse.
- Change the feeding system or physical nature of the feed, e.g. from meal to pellets. Ensure that the diet is fully balanced and adequate in trace elements.

- Mixing pigs may encourage more fighting and make the problem worse.
- Dock the tails or remove about 2 cm from the tip at birth.
- Change from flop-eared to prick-eared pigs as the incidence in the latter is often less.

Cannibalism

This is different to tail biting, although it may develop from tail biting. Cannibalism often starts by pigs attacking wounds, prolapsed rectums, sick pigs and so on.

Usually all the pigs in a pen are involved and they will chase and attack the suffering individual until it dies. The pig being attacked must be removed and the problem stopped by some environmental change for the remaining pigs such as providing straw, altering the temperature, adding chains to play with.

Gastric ulcers

These appear to increase along with greater intensification and factors suggested and examined have been infections, intoxication, nutrition, stress, gastric acidity and digestive upsets caused by diet, hormones, seasonal changes, aggression, feeding methods (especially *ad libitum* systems), housing conditions and many more.

Currently advice is to eliminate some of the causal variables to reduce the major stressors rather than any plan to treat the ulcers.

Nose rubbing and inguinal thrusting

Weaned pigs may develop the habit of nuzzling other pigs as they lie resting. It may appear as if they are navel sucking but it is more likely to be nuzzling in the flank area and along the teat line. It may be accompanied by nursing grunts. This has also been called belly nosing[23]. Persistent rubbing may cause ulcers and necrosis (destruction of the tissue).

Inguinal thrusting has been described[4] as powerful thrusting by dominant pigs (e.g. high-ranking males), again after weaning. Both these vices occur sporadically and can spread rapidly in the group and trigger off other problems like tail and ear biting. The solution is to provide some environmental diversion to reduce boredom.

Taming – handling pigs

Research[30] has shown that farms with good reproductive performance are those where a good relationship has developed between stockman and breeding stock. The pigs have clearly been tamed and have responded to good handling.

Taming is best done by quiet and gentle handling of young pigs at feeding time by laying the hand on the pig's head or back, talking gently and scratching or stroking the animal. The best areas to touch are behind the ears, shoulders, along the back and down the sides.

Pigs soon learn to enjoy this contact and will often lie down to enjoy it. Complete taming can be done in 2 or 3 weeks after which they will come when called.

Feral pigs

Feral pigs are a problem in many parts of the world and can cause considerable economic losses. They:

- Eat a wide range of farm and garden crops.
- Root up pasture and home gardens.
- Kill and eat young lambs and poultry.
- Eat garbage at camp sites and harass tourists.
- Damage fences.
- Foul water supplies.
- Cause soil erosion by their rooting.
- Are a potential reservoir for disease and its spread.

Feral pigs can multiply quickly and as they travel considerable distances, the problems they cause can be widespread. They become nocturnal when hunted or harassed but otherwise they can feed and travel during daylight or at night.

Control can be carried out by a number of methods:

- Poisoning. Approved poisons should be used and a range of baits can be obtained depending on local availability. Grain is most popular but precautions should be taken to protect non-target fauna. Advice should be obtained from local authorities.
- Shooting. Hunting pigs with or without dogs provides good sport but rarely solves the problem as the pigs become wary and all land-owners most co-operate at the same time in any extermination programme.
- Fencing. Electric fences powered by modern energisers are effective if the wires are spaced as in fig. 40.
- Trapping. This is used in areas where poisons cannot be used but its success depends on knowledge of the pigs' habits. There are many types of trap used, usually made of strong mesh material firmly anchored. They are set on regular trails or feeding areas and a wide range of baits can be used, preferably strong smelling. Pigs

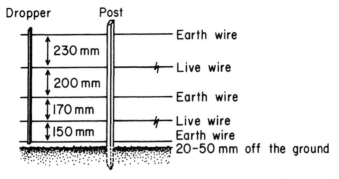

Figure 40 Design of electric fences for holding feral pigs (source: Gallagher Electronics, Hamilton, N.Z.)

caught in the trap should be regularly removed as other pigs are wary of them.

Feed can be laid around the trap and leading up to it for a few days before it is set. A trap design is shown in fig. 41[27].

Welfare
From the current information available on pigs, the following are points that should be considered in husbandry systems where there is concern for the animals' welfare:

• The feeding system must meet the nutritional needs of the animal. This is especially the case with restricted systems.

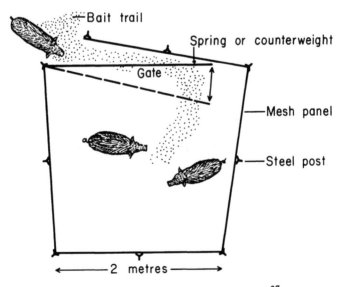

Figure 41 Design of trap for feral pigs[27]

- Housing should provide some temperature control to ensure animals are maintained within a 'comfort zone'. This zone changes with day and night.
- Uniform temperatures are not essential as long as pigs are allowed to huddle.
- Special attention should be given to the feet of pigs on wire mesh floors to prevent damage and infections.
- Feeding routines should be designed to reduce noise levels for the benefits of animals and stockmen.
- Automated systems should allow more time for regular inspection of the stock and greater social contact between stockman and animals. This should begin when animals are young.
- Pigs are easily bored and housing and management should be planned to provide for their inquisitive nature. This will prevent most vices which are the result of boredom.
- Sows and boars should be allowed proper stimulation at mating to achieve high reproductive performance.
- Sows should be given bedding before farrowing.
- Pigs should be given adequate space. They fight to obtain and maintain space.

Good stockmanship remains the goal of all systems but is especially important in large intensive units where through automation people are replaced by mechanical devices.

Further information on the production hazards in confined systems is summarised in reference 39, and the ethological demands of pigs for Europe are described in reference 6.

References

1. Allen, W.M. (1977) 'Losses of pigs due to the "acute stress syndrome" in the United Kingdom.' *N.J.F. Sympos.* Hindsgavl Castle, Denmark, 1977, 9pp.
2. Barber, R.S., Braude, R., Mitchell, K.G. and Thomas, J. (1980) 'A note on the housing of intact male and female pigs together in the same pen during the growing period to bacon weight.' *Anim. Prod.* **31**, 321–2.
3. Baxter, S.H. and Mitchell, C.D. (1977) 'Developments in floor construction in animal production.' *Vet. Ann.* **17**, 286–91.
4. Blackshaw, J.K. (1980) 'Environmental design and responses of weaned pigs.' In *Behaviour in relation to Reproduction, Management and Welfare of Farm Animals*, eds M. Wodzicka-Tomaszewska, T.N. Edey and J.J. Lynch, Univ. New England Press, N.S.W., pp. 143–5.
5. Blackshaw, J.K. (1981) 'Effect of environmental design on the growth and behaviour of weaned pigs.' *Livestock Prod. Sci.* **8**, 367–75.

6. Bogner, H. (1982) 'Ethological demands in the keeping of pigs.' *Appl. Anim. Ethol.* **8**, 301–5.

7. Booth, W.D. (1980) 'Endocrine and exocrine factors in the reproductive behaviour of the pig.' *Sympos. Zool. Soc.* London. **45**, 289–311.

8. Braude, R. (1976) 'Some problems in pig keeping.' *J. Royal Agric. Soc. (Engl)* **137**, 60–4.

9. Broadfoot, J. (1970) 'The pig (Sus) a misunderstood animal.' *J. Inst. Anim. Techn.* **21**(4), 150–57.

10. Bruce, J.M. (1981) 'Ventilation and temperature control criteria for pigs.' In *Environmental Aspects of Housing for Animal Production*, ed. J.A. Clark, London: Butterworths, pp. 197–216.

11. Buchenauer, D. (1981) 'Parameters for assessing welfare, ethological criteria.' In *The Welfare of Pigs*, ed. W. Sybesma, The Hague: Martinus Nijhoff Publishers, pp. 75–89.

12. Christenson, R.K. and Ford, J.J. (1979) 'Puberty and oestrus in confinement-reared gilts.' *J. Anim. Sci.* **49**(3), 743–51.

13. Close, W.H. (1981) 'The climatic requirements of the pig.' In *Environmental Aspects of Housing for Animal Production*, ed. J.A. Clark, London: Butterworths, pp. 149–66.

14. Cole, D.J.A. (1972) *Pig Production*. London: Butterworths, pp. 1–457.

15. English, P.R., Smith, W.J. and MacLean, A. (1977) *The Sow – Improving her Efficiency*. Farming Press Ltd: U.K., pp. 1–311.

16. Esbenshade, K.L., Singleton, W.L., Clegg, E.D. and Jones, H.W. (1979) 'Effect of housing management on reproductive development and performance of young boars.' *J. Anim. Sci.* **48**(2), 246–50.

17. Ewbank, R. (1970) 'Aggressiveness in pigs.' *Proc. Symp. Stress in Pigs*, Beerse, pp. 7–10.

18. Ewbank, R. (1976) 'Social hierarchy in suckling and fattening pigs: A review.' *Livestock Prod. Sci.* **3**, 363–72.

19. Ewbank, R. and Bryant, M.J. (1972) 'Aggressive behaviour amongst groups of domesticated pigs kept at various stocking rates.' *Anim. Behav.* **20**, 21–8.

20. Ewbank, R. and Meese, G.B. (1971) 'Aggressive behaviour in groups of domesticated pigs on removal and return of individuals.' *Anim. Prod.* **13**, 685–93.

21. Ewbank, R., Meese, G.B. and Cox, J.E. (1974) 'Individual recognition and the dominance hierarchy in the domesticated pig. The role of sight.' *Anim. Behav.* **22**, 473–80.

22. Fraser, D. (1975) 'The effect of straw on the behaviour of sows in tether stalls.' *Anim. Prod.* **21**, 59–68.

23. Fraser, D. (1978) 'Observations on the behavioural development of suckling and early-weaned piglets during the first six weeks after birth.' *Anim. Behav.* **26**, 22–30.

24. Graves, H.B., Graves, K.L. and Sherritt, G.W. (1978) 'Social behaviour and growth of pigs following mixing during the growing-finishing period.' *Appl. Anim. Ethol.* **4**, 169–80.

25. Hansen, K.E. and Curtis, S.E. (1981) 'Prepartal activity of sows in stall or pen.' *J. Anim. Sci.* **51**(2), 456–60.

26. Hansen, L.L., Halgelso, A.M. and Madsen, A. (1982) 'Behavioural results and performance of bacon pigs fed *ad libitum* from one or several self-feeders.' *Appl. Anim. Ethol.* **8**, 307–33.

27. Hart, K. (1979) 'Feral pig problems on the South Coast.' *Agric. Gaz. N.S.W.* **90**(6), 18–21.

28. Hartsock, T.G. and Graves, H.B. (1976) 'Neonatal behaviour and nutrition-related mortality in domestic swine.' *J. Anim. Sci.* **42**(1), 235–41.

29. Hemsworth, P.H., Beilharz, R.G. and Brown, W.J. (1978) 'The importance of the courting behaviour of the boar on the success of natural and artificial matings.' *Appl. Anim. Ethol.* **4**, 341–7.

30. Hemsworth, P.H., Brand, A. and Willems, P. (1981) 'The behavioural response of sows to the presence of human beings and its relation to productivity.' *Livest. Prod. Sci.* **8**, 67–74.

31. Hemsworth, P.H., Findlay, J.K. and Beilharz, R.G. (1978) 'The importance of physical contact with other pigs during rearing on the sexual behaviour of the male domestic pig.' *Anim. Prod.* **27**, 201–7.

32. Horrell, I.J. and Bennett, J. (1981) 'Disruption of teat preferences and retardation of growth following cross-fostering of one week old pigs.' *Anim. Prod.* **33**, 99–106.

33. Jensen, P. (1980) 'An ethogram of social interaction patterns in group-housed dry sows.' *Appl. Anim. Ethol.* **6**, 341–50.

34. Jeppesen, L.E. (1982) 'Teat-order in groups of piglets reared on an artificial sow. I. Formation of teat-order and influence of milk yield on teat preference.' *Appl. Anim. Ethol.* **8**, 335–45.

35. Jones, J.E.T. (1966) 'Observations on parturition in the sow. Parts I & II. The prepartum phase.' *Brit. vet. J.* **122**, 420–426, 471–8.

36. Kelley, K.W., McGlone, J.J. and Gaskins, C.T. (1980) 'Porcine aggression: Measurement and effects of crowding and fasting.' *J. Anim. Sci.* **50**(2), 336–41.

37. Kuipers, M. and Whatson, T.S. (1979) 'Sleep in piglets: an observational study.' *Appl. Anim. Ethol.* **5**, 145–51.

38. Lendfers, L.H. (1970) 'Transport stress in pigs.' *Proc. Sympos. 'Stress in Pigs'*, Beerse, pp. 57–67.

39. Ludvigsen, J.B. (1980) 'Production hazards in technically advanced confinement systems for pigs.' *Anim. Regul. Stud.* **3**, 99–103.

40. McBride, G. and Wyeth, G.S.F. (1964) 'Social behaviour of domestic animals. VI. A note on some characteristics of "runts" in pigs.' *Anim. Prod.* **6**, 249–52.

41. Meese, G.B. and Ewbank, R. (1973) 'The establishment and nature of the dominance hierarchy in the domesticated pig.' *Anim. Behav.* **21**, 326–34.

42. Moss, B.W. (1978) 'Some observations on the activity and aggressive behaviour of pigs when penned prior to slaughter.' *Appl. Anim. Ethol.* **4**, 323–39.

43. Noyes, L. (1976) 'A behavioural comparison of gnotobiotic with normal neonate pigs, indicating stress in the former.' *Appl. Anim. Ethol.* **2**, 113–21.

44. Randall, G.C.B. (1972) 'Observations on parturition in the sow.' *Vet. Rec.* **90**, 178–82.

45. Sainsbury, D. (1976) *Pig Housing*, 5th edn., Ipswich: Farming Press Ltd, pp. 1–220.

46. Signoret, J.P. (1970) 'Reproductive behaviour of pigs.' *J. Reprod. Fert. Suppl.* **11**, 105–17.

47. Signoret, J.P. (1971) 'The mating behaviour of the sow.' In *Pig production.*

Proc. 18th Easter Univ. Sch. Notts. ed D.J.A. Cole, London: Butterworths, pp. 295–313.

48. Signoret, J.P. (1980) 'Mating behaviour of the pig.' In *Behaviour in Relation to Reproduction, Management and Welfare of Farm Animals*, eds M. Wodzicka-Tomaszewska, T.N. Edey and J.J. Lynch, Armidale, N.S.W.: Univ. New England Press, pp. 75–8.

49. Sybesma, W. (1981) (ed.) *The Welfare of Pigs.* The Hague: Martinus Nijhoff Publishers, Commission of the EEC, pp. 1–334.

50. Van Putten, G. (1980) 'Objective observations on the behaviour of fattening pigs.' *Anim. Reg. Stud.* **3**, 105–18.

51. Van Putten, G. and Dammers, J. (1976) 'A comparative study of the well-being of piglets reared conventionally and in cages.' *Appl. Anim. Ethol.* **2**, 339–56.

52. Whatson, T.S. (1978) 'The development of dunging preferences in piglets.' *Appl. Anim. Ethol.* **4**, 293 (abstr).

53. Whittemore, C.T. and Fraser, D. (1974) 'The nursing and suckling behaviour of pigs. II. Vocalization of the sow in relation to suckling behaviour and milk ejection.' *Brit. vet. J.* **130**, 346–56.

54. Winstanley, M. (1979) 'A cure for stress.' *New Scient.* **85**(1182), 594–6.

55. Wood-Gush, D.G.M. and Csermely, D. (1981) 'A note on the diurnal activity of early-weaned piglets in flat-deck cages at 3 and 6 weeks of age.' *Anim. Prod.* **33**, 107–10.

General reading

Eusebio, J.A. (1980) *Pig Production in the Tropics*, London: Longmans.

MAFF (1977) *Pig Husbandry and Management*, HMSO Bull. 193.

Morris, G.L. and Curtis, S.E. (1982) 'Let the pig turn on the heat.' *Pig Int.* Jan., pp. 30–32.

Pond, W.G. and Houpt, K.A. (1978) *The Biology of the Pig*, Comstock Pub. Assoc., Cornell Univ. Press.

Thornton, K. (1974) *Practical Pig Production.* Ipswich: Farming Press Ltd, pp. 1–190.

8 Hen

The domestic hen is of special interest to animal behaviourists because commercial poultry systems have been intensified so rapidly to meet economic pressures. As a result, the hen has attracted more welfare concern than any other farm animal. Intensification however, is not new, as hens have been kept in very limited space in many cultures of the world since the dawn of history.

Behavioural definition

A hen is wary and shy, of limited ability and flexibility although it has good powers of visual discrimination. It developed in a bamboo forest habitat and its behaviour fits it for that niche. Though reluctant to fly, it uses both ground space for eating, dust bathing and nesting, and vertical space for roosting and crowing during its daily cycle of activity. Pecking and scratching are very characteristic and are the major ways hens deal with their environment and maintain their body plumage.

Their elaborate social behaviour includes responses which maintain an immediate personal space about the head when other hens are close by; reactions based on body posture and sight at intermediate distances and an elaborate but poorly understood sound communication system used when birds are further apart or hidden.

As a seasonal breeder the hen is secretive about its nesting site, lays on a 26-hour cycle up to 11–15 eggs and incubates them with only one major daily break for eating and plumage care. The male acts as an escort to and from the nest during the laying phase, and regularly mates the hen. The chicks are precocial and rapidly imprint on to the first moving object seen which would normally be the dam. They move with the dam, feeding and being brooded, as the clutch is taken around the range. This provides a daily routine for the clutch. The group remains together even for some weeks after the dam has left them and returned to the harem and tree roosting site.

The jungle fowl and feral hens

The modern hen has developed from the jungle fowl which through domestication and early selection developed such traits as leanness, aggression, activity, pecking, leg slashing, social responsiveness and particular colours. In later historic times, such as in the Roman Empire, egg and meat production were improved under systems often approaching the scale of modern operations.

Behaviour studies of wild jungle fowl of South East Asia have highlighted the birds' daily cyclical activity pattern of roosting, feeding, drinking and nesting with omnivorous feeding habits and secretive wary movement patterns. Hens were found in association with cocks, the hens' territory being about 1 ha and the cocks' about 5 ha.

Studies[43] of hens in a population that had been feral for about 100 years showed the following main features:

- The birds established roosts, about 60 m apart and with 6–30 birds/roost and 1 harem/roost.
- The amount of crowing by the males was related to status with the most crowing by the subordinate males. The dominant male acted as a suppressant to all hen fighting in his group.
- The hens nested within 45 m of water but when broody only left the nest for a short period each day.
- The broody hen and her chicks kept to themselves and threatened other hens that came within 6 m.
- Chicks start to be left by the dam at 5–6 weeks when she returned to roost in a tree. At 10–12 weeks when chicks were feathered, the hen started to threaten them.
- Chicks run ahead of the hen before weaning but behind her thereafter.
- At 16–18 weeks, the brood breaks up and adult behaviour patterns begin.

A study in Scotland compared the behaviour of a group of hens released into the wild with a group fenced off and given some domestic care. Information on nest selection, laying, brooding, care of young, feeding and movement was collected. The work highlights the importance of the strong maternal behaviour of the hen towards her spring-hatched chickens, walking over 3 km/day, walking with them 24% of the time. Their active working day lasted 16 hours and the hen initiated most of the behaviour especially feeding, tidbiting, pecking and scratching the ground. She also prevented fights between chicks. The importance of the male in organising a harem and preventing fighting was also shown and confirmed the earlier research findings. The removal of the broody hen, the male and total confinement, key features of modern poultry farming, mean that behaviour problems like severe pecking were bound to arise[39].

The hen's senses

Sight
Chick embryos respond to light as early as 17 days after the start of incubation. The hen cannot rotate its eye very much but they can see a field of 300° with a binocular field of 26°. Hens follow moving objects using the mobility of their head. Their acuity (sharpness) is good and they have good distance vision. They characteristically lift their heads before jumping and tests have shown that they can discriminate between squares and triangles and red and black dots.

Studies show that newly hatched chicks prefer to peck at blue objects rather than green or orange ones, although orange was preferred before green but not before red. Chicks when tested differentiated between a red-dyed liquid and blood. They found blood very aversive. Chicks quickly learn to avoid coloured food if it makes them ill and prefer to peck at round rather than flat objects.

Smell
Little information is documented on this. It is suggested that blood can be smelled by hens.

Hearing
Hens do not have an ear lobe but they have a well-developed ear. The calls given by a hen can range from the 250 cycles (the broody hen 'cluck') to about 3000 cycles (the distress call). Studies have shown that they can hear as high as 8000 cycles. Research[34] found that hearing in poultry covered a range from 60–11 950 Hz with highest sensitivity from 815–2000 Hz which is their normal hearing range.

Taste
Hens have about 340 taste buds mainly on the palate and floor of the oral cavity. They are rather indifferent to sugars but can detect glucose above 2½% in solution. They tolerate a range of acid and

alkaline tastes and they are sensitive to, and avoid salty foods. Whether a hen accepts or rejects a feed is related to taste, and this similarly applies to water. They can detect water temperature differences of 2.8°C. Hens will reject water that rises 5.5°C above their body temperature although they will readily drink freezing water.

Touch

Stroking, rotating and turning hens upside down will immobilise them for various periods of time. In this state, although fully alert to their sensory surroundings, they can be conditioned or gentled to humans or other frightening objects. After their return to normal stance they will show reduced fear to the conditioned object.

Touching the back of a hen will often cause it to respond by a sexual crouch, especially if it is low in social rank.

Learning ability

Hens soon learn to pull, tug, peck and scratch and their general natural activity means they will work at tasks for quite long periods. Hens have a limited ability to generalise, i.e. they stick to the task in hand and do not drift off target. They are good at visual discrimination tasks. Their limited flexibility may be an advantage in intensive conditions where they will not get bored as quickly as a species like the pig which can generalise.

In maze tests hens ranked after man, pig, dog, goat and before rat, rabbit, cat and turtle. In an ever-changing series of detour problems in a Hebb-Williams closed-field test, hens ranked after dogs, cows, sheep, pigs, goats, cats, rats and ferrets but better than pigeons, guinea pigs or possums.

Reproduction

The patterns of courtship and mating in present day domestic poultry closely parallel that of the Red Jungle fowl[60,61]. The typical responses between male and female are shown in fig. 42 re-drawn from reference 28.

The male

Males normally reach sexual maturity between 12 and 16 weeks of age. However this varies greatly with the management routine and genetic background. The feeding programme and lighting regime are of extreme importance, especially with modern strains of birds.

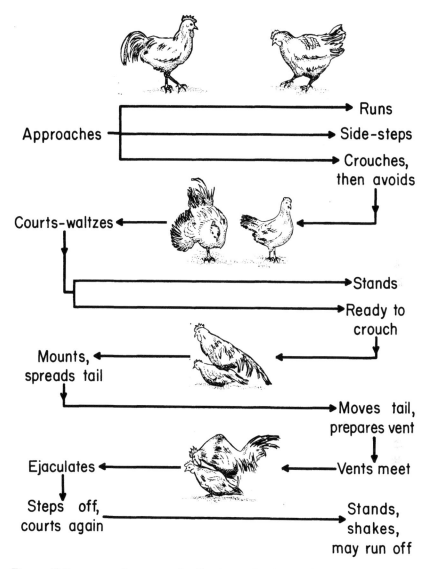

Figure 42 Sequence of responses leading to mating in hens (after Hafez (1975) — see Further reading, p. 293)

One study showed faster growth under green light although they ate more and those reared under red light had less gonadotrophin sex hormone.

Crowing by the male which increases greatly between 24 and 68 weeks has been studied and although it was not related to the bird's ability to mate, it did indicate the cock's general vigour of premating courtship and appears to be more related to aggression than sex.

The best method of scoring males for sexual activity has been debated widely. One study[52] proposed that the number of matings in eight observation periods be used, but two lines selected for this trait showed no differences in the progeny reared in single- or mixed-sex groups. Crowding does not seem to affect male survival or fertility.

Commercial success in hatcheries is measured by hatchability %, and male mating ability can have a very variable effect on this (0–74% in some studies). The following are observations from various studies:

- Previous mating experience is important to the male.
- Heavier males court less and have fewer matings than light males but they have higher sperm numbers.
- Females crouched more often for young males which did more mating than older males.
- Although high social rank males initially mate more hens than low ranked males, this advantage is short-lived.
- Most matings occurred 15 hours after lights on[11] which generally means a peak (80%) between 4 p.m. and 6 p.m. depending on lights-on time.
- Male-female ratios of 1:5, 1:10, 1:12 or 3:40 were equally successful.
- Change in light pattern increases or stimulates male sperm production, though normal semen production occurs with standard light levels.
- Depriving males of water for 48 hours will reduce semen production for up to 6 weeks.
- Sperm quality is reduced by high levels of iodine (5000 p.p.m.) and mouldy feeds.
- Sperm production is improved by low temperatures.
- In sexually active males, massage will not produce semen as it will in low-activity cocks.
- Males will mate from 1 to 53 times/day.
- Males kept together will tread (i.e. mount) each other.
- Courtship. Caged males spend more time waltzing than colony males, the latter moving up to hens from the rear, rather than the side.
- Males in floor pens did more fighting, courting and mating than those in colony cages. Females avoided males more in the floor pens.
- Bouncing floors upset males more than firm wire floors.

The capon

A capon is a castrated male. Castration used to be done by surgical procedures removing the testicles through an incision in the body wall. It is now done by oestrogenic hormones implanted under the skin of the neck. This arrests testicle growth, combs and wattles are smaller and birds appear more docile with no sex drive. However, hormonal implantation is banned in some countries so legislation should be checked.

The female

In free-living systems, young females (pullets) show mating behaviour from as early as 18 weeks of age, and all hens, regardless of age show characteristic mating behaviour 4–8 days before the new laying season starts.

Early research[27] showed that the status of the hen greatly affected her mating behaviour with low-status birds crouching more readily and being mated more than high-status hens. Low-status hens were also lighter.

Environment, especially feeding, lighting pattern and strain have important effects on sexual maturity.

Broodiness

The length of time a hen remains broody varies greatly and modern laying strains have been selected effectively against broodiness. Thus the best broody strains are the older non-hybrids. A good test of the degree of broodiness is to give the hen some dummy eggs to sit on and watch to see how tight she sits and how she resists being disturbed. The main practical points in managing a broody hen are:

- Make the nest from a hollowed-out turf which is damped during sitting to provide humidity. Cover this with dry straw.
- Treat the broody for external parasites before sitting.
- Let her sit on a few dummy eggs for a day or so then replace by 10–15 eggs depending on her size.
- Allow the hen off the nest each day to feed, drink, exercise and defecate. Avoid constipation in the broody hen.
- If she does not leave the nest, lift her off making sure there are no eggs under her wings.
- Make sure she sits on the nest until all the fertile eggs are hatched.

Pre-hatching and hatching

Embryos transferred into glass shells have shown that pre-hatching activity goes in bursts. The chick has its head usually under its right wing and with each burst of activity the limbs extend and flex and then return to their normal positions. When due to hatch, the body is up against the shell and during this activity the beak performs the cracking of the shell. Limb movements are in a set order which causes a counter clockwise movement of the embryo in the shell. No activity occurs when the chick is in the sleep phase[2]. Incubation details are reviewed in reference 41.

Immediately on hatching there is rapid development with extension of the wings and flexing of the hind limbs. This causes a crawling action which helps the chick to stand upright and begin to take a few steps. The neck can be held erect long before the chick can stand up.

Chicks hatched early from incubation (e.g. day 19) can suffer from dehydration and dried-out yolk sac, and chicks hatched late (e.g. day 21) can suffer from leg problems. Peak hatchings occur at 20½ days after setting and farmers aim to have their chicks delivered 24 hours after hatching, well before they need water and food.

Brooding temperatures for incubator hatched chicks are 30°C dropping to 19.5°C after 5 weeks.

A hen will accept strange chicks 2–5 days after hatching if they are the same colour as her own. Thereafter she may kill them.

Euthanasia of chicks

Many methods of killing cull chicks have been tried but according is Jaksch[38] none is fully successful. The methods available are:

- Manual – neck dislocation.
- Mechanical – neck dislocation, decapitation, or high speed crushing.
- Suffocation or decompression by withdrawal of oxygen.
- Gassing with nitrogen or carbon dioxide.
- Electrocution.

Manual methods are labour intensive and high speed mills killing 500 chicks/minute are for large-scale operations. Decompression is expensive while nitrogen and carbon dioxide take up to 15 minutes for death. Carbon dioxide alone seems the best at present with un-consciousness after 10–15 seconds followed by death. The chicks must stay in the gas chamber for 30 minutes. None of these methods meets all the humane criteria to satisfy all concerned people.

Social Behaviour

Vocal communication

Being a social bird from an open forest habitat, the hen uses a wide range of clucks, cackles, chirps and cries to keep in contact with other hens. One study[18] classified 12 chick calls and 22 adult calls. The ones heard most often are food calls, predator alarms, pre- and post-laying calls and rooster crowing. Others are more specific. One classification describes vocal calls related to fear and predators; physiological calls to do with brooding, feeding, contact and pleasure; and then signals related to fear or pain, frustration, fighting and crowing.

There is currently little uniformity of agreement among workers on the role and range of vocal communications in hens. There is however a well-recognised daily pattern of crowing near dawn followed by feeding calls, egg-laying calls and finally roosting calls. Chicken distress calls immediately draw the attention of their broody mother.

Noise levels created by large groups of hens can be high. One study showed that laying hens were affected by noise levels above 83 dB while higher levels induced fear and panic. Noise ranges from 72–87 dB at normal times, 73–100 dB at feeding and 75–85 dB during egg laying. These noise levels can also affect the health of attendants.

Body posture

When hens can see each other, they communicate by body postures such as head up or down, tail up or down, feathers spread or not. The tail is especially important and studies of feral birds[43] showed that they stood upright with tail erect with wings diamond-pointed almost vertically down. This they called 'wing-down alert'. In modern intensive systems the practical value of body postures is of little importance.

Suggestions have been made that visual models of an alpha male could perhaps be used to help hens settle and lay, or specific signs could be used to get shy eaters to feed. More research is needed in these areas.

Individual recognition

Research[7,28] has shown that the areas concerned in individual re- cognition between hens were the combination of comb, head and wattle. Single elements were more difficult for hens to recognise but the comb was the simplest. Colour changes to plumage were noticed with intense colours more easily seen than pale colours. Only abrupt and very dramatic changes cause a hen to be treated as a stranger.

Pecking and peck order

Pecking is a recognised species-specific behaviour of hens. Hens peck to get out of the shell, to feed and drink, obtain space and recognition, to mate and many other functions.

Hens maintain a personal space around their heads and they keep a distance from each other holding their heads at an angle and maintaining a specific body orientation. If a direct head-to-head stance is taken, then pecking will ensue.

However, the main purpose of pecking is in eating and it is a very precisely tuned movement of the head and neck, the food being first picked up and then in a further movement of the head it is swallowed. Incorrect timing of these actions would severely jar the neck. The birds' binocular vision is important in judging the distance to peck with the eye membrane closed at the movement of impact. A filmed study of pecking movements has been made[37].

The relationship of body stance and head position is important during pecking. Attacks include threats in which one bird lifts its head above the level of the other bird's head, then pecks the comb, head, neck or nape, wattles and then chases the subordinate. If two birds face up to each other to fight they peck, kick with their feet and slash with their spurs. Submission is shown by crouching or escaping. There is a low incidence of this fighting behaviour out of doors but it increases as stocking densities move towards 460 cm^2/bird.

In free-living hens pecking is greatly reduced when males are present. In indoor-intensive conditions, trials have shown that the influence of the male to reduce pecking was greatest when hens were on the floor rather than in colony cages. Recorded male calls had no effect.

Pecking is often greatest in adolescent hens and observations have shown the incidence to be 30–50% greater on floors than in cages. These hens gave 70% more pecks when males were absent from the group. Attempts to substitute high-rank hens with young birds had limited success as these older hens also pecked at others while they were feeding, a practice not done by the male.

The famous technical bulletin of Guhl[26] set out the implications of peck orders in fowls and these are fully discussed in other texts (see references 10, 22, 61). The following are some practical points to stabilise a peck order quickly so that a minimum egg production loss occurs:

- Form new groups of hens by mixing them before production starts.
- Do not revolve birds around groups.
- Provide plenty of feed and water points and sufficient floor area when the flock is settling.
- If two groups have to be mixed, put equal numbers of each subgroup together.
- Make sure males have been together in a group before mixing with the hens.
- Putting a male among hens will reduce the level of pecking.

Much of the pecking in caged birds occurs during feeding bouts, and depends on the feeder space and number of birds/cage.

Dust bathing

Dust bathing is claimed[57] to be a behavioural need of the hen, and its function is assumed to be to get rid of ectoparasites, and align the feathers. It may also be an alternative to aggression although this is very speculative. Normally free-living hens spend their time dispersed except for dust bathing when they crowd together.

Dust bathing is also thought to occur as a type of vacuum activity seen in birds that have been released from cages after at least 100 hours without a bath, and in birds that have had previous experience of the practice.

Dust bathing usually starts with pecking into an area of dry dust, squatting in it, turning and raising dust into the feathers and then shaking it out again. Whether dust bathing is a basic need of hens is still in doubt.

Preening and feather care

Researchers[58] who studied grooming in the hen described it as those actions related to maintaining body surface including preening with the beak, scratching with the foot, dust bathing and oiling. Preening in chicks has been described and in adult birds there is a strong relationship between the amount of grooming and the number of lice removed. Hens are more efficient groomers than cocks and they also groomed more and used their oil (urophygial) gland more often than cocks. In preening the hen raises the feathers and cleans them by stroking with the beak and nibbling them.

Other comfort behaviour

Bill wiping is seen after birds have been eating wet mash. The bird wipes one side of its beak on the ground and in one continuous movement wipes the other side. Unilateral stretching of the leg and wing is a comfort behaviour along with wing flapping.

Roosting or perching

The wild jungle fowl roosts up off the ground in a communal roost in the centre of its territory. Some recent studies[20,21] of perching of intensively housed hens have shown that some birds did not perch at all and even repeated exposure to perches did not change their behaviour. Birds did not avoid wire mesh as a perch and they varied in their choice of high and low perches.

Rearing

Precocity

Chicks are very precocious and are able to 'get up and go' very soon after hatching. In the wild, survival depends greatly on rapid bonding between the hen and her chicks but an important question not fully answered is the effect which artificial hatching and rearing has on subsequent hen behaviour in such traits as feeding and nesting.

On the other hand, the chick's precocity allows it to adapt to rearing without a broody hen in intensive systems. Chickens reared in fully sterile conditions did not develop an effective gut flora which they would normally pick up after the tidbiting of the hen to drops of her saliva.

Bonding

Chicks will approach and follow a moving object about 1 hour after hatching and zoologists and psychologists have shown wide interest in how hen and chicks become bonded. Chicks have been imprinted on to moving balloons, revolving coloured wheels or any object that stands in marked contrast to the background. The period for imprinting and bonding is from 9 to 20 hours after hatching. Some pre-hatching auditory imprinting has been reported in some precocial birds, but its importance in poultry is not known.

The hen 'clucks' and the chick 'peeps'. The more the hen clucks, the fewer peeps the chicks make. Chicks moving with their dam utter contentment twitters or distress cheeps. If the hen stops and calls, the chicks remain stationary. If she is too far away the chicks peep and she goes to brood them. If the call of a foreign mother is played, the chicks stay still for longer and peep less often. This reduces the chance of their being attacked by a strange hen. The length and loudness of the call control chick behaviour while the sound frequency leads to recognition of their own parent's voice. Chicks feed freely in the presence of their own dam's call while alien calls will halt feeding altogether.

As chicks cannot recognise each other much before 10 days of age, the hen and the calling system keeps the brood together and prevents aggression among the chicks. The hen also teaches the chicks to react to food and to predators. Chicks normally show fear 33–36 hours after hatching, but this is extended if they are kept in isolation. The experience of communal feeding needs to be achieved before 3 days post hatching.

Other behaviour

Chicks are very active and when running they extend their wings and flap them for use as brakes. They jump onto feeders but do not perch till 4–6 weeks of age. They stretch in a very precise way with a wing and leg on one side stretched out pointing to the rear with the wing primary feathers displayed. Chicks spend a lot of time chasing and if they turn, face up and stare at each other, this can lead to regular fights by 2 weeks of age. These fights are only between two birds at a time (usually males) where they grab at neck feathers and pull the adversary to the ground. Pecking and feather pulling clearly induces pain from day 13 and by day 18, weaker chicks can be pulled down and trodden.

Dust bathing starts at day 3 and is a copied activity. Preening of wing and breast feathers may start at day 2 but no preen gland is used until day 14. These preening spells may last up to 4 minutes. Chicks will start picking at toes by 10 days of age, and especially if conditions are hot, dry, and too bright, they will pick at wing and tail feathers until they bleed as well as pecking at pasted up vents. They will pick at any bright object in the litter. This may include nails and staples which can cause death if eaten.

Litter scratching is a very stereotyped action from day 2, best described as a scratch with the right leg, then two with the left, one with the right and so on, while the litter is flicked over with the beak. The whole sequence takes about 15 seconds and occurs in the best-lit areas first. If the environment is too cold, litter may be eaten and the gizzard impacted.

A major study[17] of behaviour in chicks up to 10 weeks of age recorded many changes. Resting was high near hatching, declined till 3½ weeks old, rose to 8 weeks and then declined again. Females rested more than males. Stretching legs and wings increased (particularly in males) to a peak at 4–7 weeks old and then declined. Scratching increased in the first week, declined to almost nothing at 7 weeks then reappeared at 9 weeks of age. Preening increased with feather growth while running about decreased with age. Frolicking also increased up to week 4 and then was seen to decline as sparring started and reached a peak at 5 weeks old. Actual agonistic encounters replaced sparring by week 7 and this pecking reached a peak at 8 weeks by which time a clear peck order was established.

Temperature

The chick's first response after hatching is to seek warmth and cover, and they are very vulnerable to draughts. The body temperature of the day-old chick is 38.6±0.6°C rising to 40.6±0.5°C by day 9. If heating is inadequate, chicks huddle together in a semicircle facing outwards and can easily smother in corners away from draughts or in sunny spots. When too hot they disperse away from the heat source (fig. 43). An optimal brooder temperature is 30°C for the first 3–4 weeks reducing to 21°C for subsequent weeks. As feathering starts the brooder temperature may be further

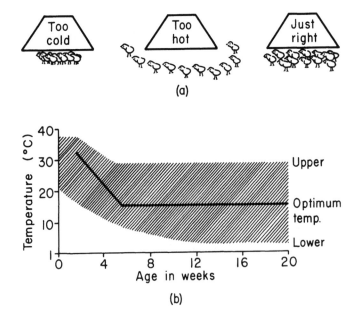

Figure 43 Brooding house temperatures. (a) These can be adjusted from watching chick behaviour; (b) optimal temperature can be changed as chicks grow (source: S. Australia Dept Agric. Bull.)

reduced, using the chick's reactions as a guide. Chicks will not move out into a colder temperature for food before 5–7 days. Low rearing temperature (e.g. 19.5°C) will slow growth, cause earlier feathering but increase the size of the heart, thyroid and adrenals compared to chicks reared at 30°C.

Drinking

As chicks dehydrate quickly at the high temperatures of rearing, it is vital that they find water quickly and learn to drink. Drinking often starts by the chick pecking at a bubble. Some water movement is important as even standing in still water may not trigger them to drink. One method is to lay paper on the floor on which food and water is placed so that they find it through their pecking. Remove the paper after 1 week because of disease risks. Be guided by the behaviour of the smallest chicks. If they have found water, then it can be assumed that the others have. Chicks should be given 24 hours of light at this time to allow them to find food and water. Some chicks may peck at moist droppings which stick on their beaks. By 3½ weeks the chick may in a minute-long drinking session, drink 11 times with a few seconds pause between each. The chick drinks by placing the beak in the water and scooping up a beakful which it then swallows by holding its head horizontal or slightly raised.

Water should be available at all times and offered at least 4 hours before feed[47]. Information is still needed on the best design of drinkers to start the birds off early with good drinking habits. Studies[45,48] have shown that many chicks just approach the drinker but never use it. Table 7 shows the percentage of chicks involved in drinking activities depending on the shape of the drinker.

The low percentage of chicks drinking is obvious as is the variation in drinker design. On some drinkers the lips are above an optimal

Table 7 Percentage of chicks involved in drinking activities (from reference 45)

Type of drinker	Stayed at drinker	% those that stayed touched drinker	drank	% of total drinking
Bell	73	37	59	16
Bottle	62	54	69	23
Trough	91	59	67	36

height of 6–7 cm above the ground. Troughs encourage social facilitation in drinking. Roosting on some models of bell drinkers may cause automatic water shut-off. Much more behaviour research is needed on drinker design for shape, colour, placement, space/bird and so on.

Eating

Food recognition by the chick is complex. Chicks will peck indiscriminately at various objects in their environment such as sand grains, shiny objects, other chicks and so on, and almost by trial and error discover food. If reared by a broody hen, the chicks' feeding problems are greatly reduced by her tidbiting for them, encouraging them with vocal calls to eat the food items she indicates.

The use of glitter tape around the troughs is sometimes used to attract chicks to food as is the use of glass marbles in the troughs or making starter foods into three-dimensional crumbs or crumbles to encourage pecking. Egg trays make good feeders to start chickens eating. In a French study using a 10-hour light day regime, chicks up to 3 days of age only spent about a third of their time eating but by ten days old this changes to nearer 60%. Observed chicks ate thirty-four times on day 3 and 123 times on day 10. In further studies, chicks ate solid feed 30–55 times/day and they drank water 22–45 times/day. As age increases the number of feeding times dropped but total feed eaten increased.

Beak trimming

Beak trimming is done by removing some of the top (usually a third) and bottom beak with special precision cauterising blades (see fig. 44). The bird is thus able to eat but cannot peck so it is a common technique to prevent cannibalism in birds of all ages in all types of systems. It is accepted as a labour-intensive necessity by many producers.

Figure 44 Correctly trimmed beak of chick. This reduces pecking but will not hinder feeding (source: Peter Josland)

Chicks are usually done at 5 days of age and it can be repeated at any stage when severe outbreaks of cannibalism occur. However, if properly done at 5–7 days, further trimming should not be necessary. Sometimes beaks are touched up at 16–18 weeks. Care is needed to have the blade sharp, at the correct heat, and to remove the correct amount of beak with least stress to the birds. Poor trimming can set a flock back by 2–3 weeks because of burnt tongues and nostrils which are slow to heal. This results in reduced feed and water consumption, loss of body weight and lowered production.

Studies showed that pecks from trimmed birds were ignored to a much greater extent (58% more) than pecks from intact birds. These trimmed birds showed higher levels of pecking but social rankings were not changed by trimming. Indeed, they had to peck even more to keep their place although less damage was done. There is no doubt that poor beak trimming has more disavantages than cannibalism[3,29]. In well-run cage systems where cannibalism is rare, beak trimming is not needed.

Handling

The influence of early handling (before 3 days of age) termed Tender, Gentle, Care or TGC, has been studied to see its effect on behaviour and performance in later life[24,25]. The handled group showed better growth, improved resistance to *E. coli* and an improved ability in later life to accept stress. In practice it seems worthwhile for stockmen to have a deliberate policy of handling birds gently, speaking to them quietly and doing some feeding from the hand. Frequent visits by the birds' attendants in this early stage is very important. Large intensive units need to exploit TGC in their systems.

Production – eggs

Nesting

In free-living hens (e.g. on free range or litter) selection of the nest site is done with great care, often in association with the male. Nesting is characterised by secrecy and careful nest concealment.

Several major studies have focussed on behaviour surrounding egg laying and it is clear that nesting activity and ovulation are related so that a good nest favours high production. The hen normally moves through four phases, as shown in fig. 45.

- Seeking out a place to lay. This can be very protracted as she becomes restless, paces about giving the pre-laying call and showing

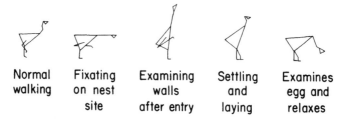

| Normal
walking | Fixating
on nest
site | Examining
walls
after entry | Settling
and
laying | Examines
egg and
relaxes |

Figure 45 Postures shown by hens in the period immediately before laying[60]

characteristic body postures[60]. In a deep-litter laying house she examines the walls and corners.

- Inspection of a number of possible sites between feeding, preening or sleeping and finally pushing into one of them. She continues to examine the nest, lifting her legs with care and keeping her neck horizontal.
- Settling, squatting, making a nest hole by crouching and rotating several times using her keel bone to shape the nest. Then she sits and lays, often standing to expel the egg.
- After laying, she may examine the egg with her beak then rises and returns to the flock and cackles.

In feral hens, the cackle brings the male to her and she is often re-mated at this time. Hens spend most time in nest selection and it appears that this may be more important than the nest itself.

Nesting in cages

In the limited space of the battery cage, the same sequences are adopted by the hens but their expression is modified.

The hen searches the cage, pushing other hens away. She may creep between the legs of other hens, sometimes up to 100 times, before she settles. She spends laying time putting her head between the wires, pushing and often being pecked by birds in the next cage. Intention movements to fly may be shown by tail extended and wings slightly raised. She may even try to climb the cage. Then suddenly she squats and lays. Breathing rate is high and her calls are generally weak. Eggs are often laid in the same area of the cage.

Research[23] shows that the caged hen paces to fill up the time normally spent in the pre-laying behaviour seen in the free-living bird. Pacing varies between strains from 100–2600 paces before laying. Birds were recorded as spending their time in resting 55%, eating 21%, laying 17% and drinking 7%.

One researcher[6] claimed that the pacing of a caged hen does not compensate for the lack of a proper nesting site. However against this must be put the advantages of battery cages.

Nest design

In the Jungle fowl, nesting is mainly ground-bound and in a bamboo habitat bird, leaf shadow patterns and semi-darkness are probably important.

The main aim in free-living hens is to get them to lay in nest boxes where clean high quality eggs should result, and to prevent them laying on the floor.

In general one nest is needed for every four pullets and these must be provided by 18 weeks of age, i.e. 2 weeks before sexual maturity. Nests should be clean, have ample litter, such as straw, shavings, sawdust, wood fibre, rice hulls, almond shell, silica pellets, fir, cork and so on, depending on the local supply. Birds should be prevented from roosting in or on top of nesting boxes or sleeping in them.

The nest box should be large enough for the bird to turn round in and some studies have shown a bird's preference for a triangle-shaped entrance rather than a square one. Attempts have been made to associate a colour with a nest box but the practical problems remain to be solved. The aim is certainly to attract birds into the nest to lay, as although catching and confining them to nest boxes is effective it is far too labour-intensive. Nest boxes are best sited 450–500 mm above the litter level.

Leaving dummy eggs in nests is an old trick to attract hens to lay but it seems that the hen is probably more attracted by the nesting material than the eggs. A balance must be struck between nest access which if too difficult leads to floor-laid eggs, and if too easy encourages other activities in the nest box as well.

Single or communal nests

Choice of nest box layout depends on the system, with ease of egg collection being a major concern. Communal nests can cause problems as shy, low-ranking hens may not push their way into the box. They become more persistent as the laying urge increases. Hens that are sitting tight are difficult to move, especially if they are going broody.

One study[40] identified two types of hens, solitary nesters and social nesters. The solitary hens sought out empty nests and evicted other hens if all nests were full. It would seem advisable to have both single and multiple nests available in free-living poultry systems.

Communal nests of 60 x 200 cm accommodate fifty birds, and whatever kind of nest is used, there should be a rail or perch along the front on which birds can sit and wait, and inspect the nest and its occupants. However, regardless of the number of nests or their type, some will not be used and crowding will occur in others.

In situations where egg records are needed for individual birds, a trap nest is used. This is a single nest in which a slide drops down after the bird goes inside. The slide mechanism is triggered by the bird's back and tail. She cannot get out until released when her number is recorded on the egg and the record sheet. These trap nests cause no problems to the birds which readily adapt to being trapped in.

Eggs

The average brood size from Jungle fowl in a zoo park was six offspring with clutches of eggs ranging from two to ten. Hens on free range average about 220, with deep litter 245 and batteries about 260 eggs/year. Research has shown[62] that there is a 5–20% loss from ovulations that are not dropped into the oviduct but are absorbed into the body cavity. When eggs are displaced from the nest, the bird normally uses the beak to retrieve them. One bird was observed to move an egg 1.5 m.

Cracked eggs

Cracked eggs can be a major problem. Cracking occurs when the egg hits the floor after being laid, and the risks are clearly greater on wire floors than on litter. The floor mass, the slope, height of the drop when laid (which varies with the bird) and the angle the egg is presented from the cloaca are all important factors.

In cages, more cracked eggs occur on heavy gauge wire floors, recommended by Brambell[4], although preference tests showed birds preferred finer gauge wire with more foot support.

Floor laying

Eggs laid on the floor are a major economic loss to poultry producers, and as the habit is hard to cure once started, prevention must be given priority. Ground laying is natural in the wild hen so the environment must be altered to get it to lay above the ground where it normally roosts. Some practical suggestions are:

- Start with the nest boxes on the floor and slowly raise them.

- Don't make the floor litter too deep. Have 7–8 cm rather than 15 cm of litter.
- Pick up all floor eggs and leave some real eggs or some dummy eggs in the nests.
- Direct access from roosts to nests will allow shy birds in to lay.
- Make the nests attractive with plenty of litter, regularly renewed.
- Keep the nests in semi darkness. Make sure they are vermin-proof.
- Try to eliminate dark corners in the pen.

Temperature and humidity

The control of the birds' environment in modern intensive systems has become a specialised study and reference should be made to the appropriate sources for details[56]. Temperature and humidity dictate ventilation rates as do stocking densities.

Pigs indicate a preference for a diurnal temperature range and this may also apply to poultry with the potential for large energy savings. Research is warranted in this area.

Eating pattern

Feed intake is high well before egg laying and then drops 2–3 hours before lay, increasing rapidly after that. In one study where free-choice feeding was offered, birds were observed to select a better diet (i.e. more calcium, less energy and protein) than the proprietary one offered and produced more eggs with a 7% reduced intake.

As feed is the major item of cost (70–80%) in poultry production, ways to prevent feed wastage are given high priority as well as achieving high intakes by maintaining good palatability. Readers should consult the specialist texts.

Lighting

Poultry producers can control the sexual development of laying birds and the subsequent egg production by modifying the lighting pattern. The change in the pattern (e.g. whether step-up or step-down) and the light intensity is important.

Birds seem to treat the longest dark period as night and the resumption of light as dawn. Research has examined repeated short interruptions using dimmers instead of abrupt on-off periods. The whole subject of lighting is reviewed in reference 44.

Moving from a conventional 24-hour lighting schedule to a 28-hour may mean less eggs but these eggs are heavier and thicker shelled so fewer cracks occur. This is important for hens late in lay. A 28-hour cycle means that the working week is reduced from 7 to

6 days with a 1-hour light period on Sundays. It seems that laying birds need no more than 17 hours of light for optimal production.

Stocking density

Stocking density along with feed conversion efficiency are the keys to good returns on invested capital in a poultry enterprise. It is stocking density that has received the greatest attention in the welfare debate. Information[35] from well-designed studies reviewed in 1975 highlighted three separate questions to be answered:

- What is best for the bird?
- What conditions give maximum output?
- What colony size and space give best economic returns?

The research results showed that decreased area/bird reduced egg production, lowered body weight and increased mortality. Increasing colony size depressed egg production, raised food consumption and increased mortality. These effects are independent and additive so that you can get one or all of them at once.

As a general rule overall colony sizes should be kept small and when profit margins are low, the area/bird can be increased and decreased when profits rise again. This would require cages of a more flexible design and may not be practical.

Water

Water is critical for all laying birds. Birds may reject very hard water (over 1000 p.p.m. of calcium) which will also clog up drinkers. Zinc levels (over 2300 p.p.m.) will also depress water intake as will high nitrate levels (over 300 p.p.m.). There are other potential water problems caused by chemicals and algal growths.

Moulting

Moulting or loss of feathers is a normal process in the bird and usually occurs in hens when egg laying stops at the end of the season. A new generation of feathers grows and pushes the old ones out, this taking about 8–12 weeks to complete. Up to 25% of birds may be culled before their second production year when egg number/bird is 20% lower, egg size is larger but quality and shell texture are lower.

In some systems, there may be economic reasons why birds need to be kept on for a longer laying period, and to avoid the long wait during a natural moult a shorter induced moult is brought about by reducing feed, water intake and reducing light. Anti-ovulatory drugs may also be used as well as mineral imbalance in the diet.

The need is to extend the bird's normal 'pause'. A pause is a natural interval of 2–3 days after the hen has laid one egg/day for 6 days. The forced-moult technique extends this pause. There is debate as to whether an extended pause only is needed or whether a pause plus moult is better.

There is little hen–hen pecking during the moult but it increases during the recovery period, especially around food hoppers. The hens that pecked a lot also showed a compensatory (higher) feed intake. However hens have been shown to tolerate some force moulting regimes with no serious problems[33].

Production – meat

The modern broiler chicken is probably the best example of how an animal can be bred, fed and housed to meet very specific market requirements, under intensive competition with prime emphasis on efficiency. The modern broiler converts 1.95–2.20 kg of feed into 1 kg of body weight from which 0.7 kg of carcass is produced.

A good broiler has to increase its birth weight forty to fifty times before slaughter at forty-two days. Behavioural problems that arise are inevitably expressed in poor growth rates, worsening feed conversion and increases in mortality, all of which can be regularly and easily recorded in good management systems.

The following are some management points that are important for good results from meat chickens:

- Adequate drinkers are essential and should be regularly checked.
- Good feeder design is important to allow the smallest bird to eat, prevent wastage and blockages regardless of type of feed[19].
- Birds should have access to food and water within about 2 m of their area in the deep litter shed, and there should be sufficient access to allow low-dominance and shy birds to feed.
- Physical form of the feed affects intake, such as whether it is hard or soft, meal, crumbs or pellets, and the size of the pellets. Removing stale food and regular topping up of feeders is important whether manual or mechanical.
- Checks on all mechanical equipment and failsafe mechanisms with operator warnings are basic to any production plant.
- Ventilation. High levels of ammonia and carbon dioxide can cause serious problems. The atmosphere should be dry and temperatures optimal. Consult appropriate manuals.

- Stocking density and crowding. Stocking rates recommendations are given in specialist texts. Examples of a controlled environment is $17-22$ birds/m² to a maximum age which is equivalent to 34.2 kg/m². In non-controlled this would reduce to $9-12$ birds/m².
- Lighting. Details of environmental specifications are given in reference 9.

Catching, bruising and transport

Bruising in broilers is a major cause of economic loss in an enterprise where margins/bird are small. Bruising from commercial houses can reach 3% or more. Much of the bruising (80%) occurs in the catching process which is done at night or early in the morning by teams of workers who may have worked during the day and consequently become tired. The birds are held under low light intensity which again is not a good working environment for the workers. Catching requires skill and care.

Experience shows that bruising is greater on warm than cold nights, males bruise more easily than females and heavier birds bruise more easily than lighter birds. The caging of badly caught and severely stressed birds, together with the shock of daylight and a draughty, noisy truck journey can cause injuries, haemorrhages, collapse and suffocation.

Drugs may be given in the drinking water some hours before catching to partially sedate the birds and assist capture. However regulations should be checked as the practice may be illegal in some countries for fear of residues left in the carcass.

Firm, quiet handling using wire frames to confine the birds and prevent bruising is best. Birds may be starved $6-12$ hours before catching.

Slaughter and stress

There is a continuing need for more information about poultry slaughter so that killing procedures may be improved. Most broiler-processing plants deal with about $1000-6000$ birds an hour on a conveyer chain system from which the birds are suspended by their legs when still alive. This chain moves at about 10 m/minute past a killing point.

High voltage stunning is the usual practice. This has to overcome the resistance of the head and comb and can be dangerous to the human operator. The bird's resultant severe muscular contractions may fracture bones or crush the sternum and bleeding may be inhibited. Low voltage systems overcome some of these problems and an electrified water bath system can be used. One worker[51] suggested that not all birds show signs of unconsciousness when operator hand-held devices are used. Gassing as a means of killing may have potential which needs further research.

Housing systems and behaviour

Poultry production over recent years has seen large changes in housing systems. Considerable interest has always been shown by producers in comparisons between these systems but such a practice is of little value. The housing and production system must be examined for itself as to whether it meets the needs of the bird and allows high, sustained and economic performance. The philosophy of McBride must be the ultimate aim – to fit systems to birds and not birds to systems. Hann[30] reviews the advantages of various raising systems and attempts to rate them for their effectiveness.

Free range
This is a traditional system for egg production which still today has the vision of 'farm fresh eggs' associated with it, where hens can wander at will over green pastures with no environmental restrictions. There are however many drawbacks to this system[49,50], and in Britain only about 2% of poultry are now run outdoors. Some of the disadvantages are:

- Little protection from the weather.
- Greater labour requirement for feeding, egg collection and moving houses.
- More food required/dozen eggs produced.
- Greater food wastage.
- The birds' diet may be grossly imbalanced.
- Dirty eggs and fouled pasture.
- Lowered egg quality. Stale eggs and offensive flavours.
- Birds run greater risk of disease from soil-borne infections.
- Greater incidence of internal and external parasites.
- Risks from predators.
- Poor working environment for staff.

The system is still likely to remain because of the comparatively low capital cost incurred and its suitability to small-scale production. Some producers can obtain a consumer's premium for free-range eggs,

Behavioural problems are few on this system because of the unrestricted space given to the birds which allow low-ranking birds to escape molestation. However, feather pecking can rise to such levels where prevention such as beak trimming may be needed. Egg eating may also be a problem, often because it is difficult to observe and find out the problem birds. Predators also take their toll.

Slatted and wire floor (aviary systems)

Hens are kept on totally slatted or wire floors where very high densities can be kept, e.g. up to 0.09 m²/bird. Nest boxes are accessible from outside without going into the house among the birds.

The system tries to find a compromise between cages and deep litter. The perching area is a main feature of this system, and is built over a droppings pit which need only be cleaned out once a year.

The critical aspects for behavioural problems in these houses are to ensure that nests are placed near to the perching area, that eggs can be gathered easily, and that the ventilation system can handle the foul air from the dropping pit. Floor eggs can be a major problem. Long, narrow pens are good to prevent a lot of interaction between the birds but panic and bruising can arise when inspection takes place by the attendant.

Deep litter

This is where birds of all types are kept on litter at various stocking densities with partial or full environmental control.

Behavioural problems in deep litter usually seen as a drop in egg production, include increased aggression (feather pecking), mortality and feed wastage. Floor laying may also be a problem. To investigate these problems the following points should be checked:

- Temperature, humidity and ventilation.
- Feed access, wastage and diet quality and quantity. Physical form of diet.
- Type and condition of the litter. Is it wet or dry?
- The lighting pattern and intensity. Check time clocks.
- Stocking density. Check this as modification may be essential.
- Resort to beak trimming and spectacles only after the problem of aggression has been fully investigated.

Birds in large open houses keep to specific areas and do not range widely through the house. Social orders are set up within these territories, so although they are free to move, there are strict social limits on their mobility. The risks of attack are present for birds that move out of their established territory. Observations have shown that the area over which bird's move varies greatly between houses, some being much larger than others. Thus the notion of individual or personal space is more acceptable in birds on litter than that of territory or home range.

Straw yards

These are similar to deep-litter systems where deep straw is used instead of litter. They do not have full environmental control. Some people are now advocating a return to this old system and argue that it is commercially viable and meets most welfare needs. No behaviour studies of birds in straw yards are reported.

Battery cages

Cage designs have been strongly affected by economic pressures on producers, as well as the concern of welfarists.

The advantages of cages for laying birds are:

- They allow provision of a completely balanced diet.
- It is easier to maintain an optimal climate.
- Egg eating is less likely.
- Inspection of birds is easier.
- Catching birds is easier.
- Parasites and diseases should be easier to control. This may not be true in poor environments.
- High production of high quality eggs is easier to achieve from strains of bird genetically selected for this environment.
- Birds rest more, pecking is reduced and cannibalism is less likely. As a consequence at the end of production, some heavier strains of caged birds are of better body weight which has marketing advantages. However bruising and scratching can occur with poor management.

It is often difficult to determine when a 'multi-bird' cage becomes a 'colony' cage. One suggestion is that the dividing point should be fixed at eight birds/cage. It appears from current studies that about 460 cm²/bird is an economic optimum but that legislation favours 500 cm²/bird in Britain and 600 cm²/bird in Denmark.

Behaviour in cages

The most useful work has been to examine the responses of birds in conventional cages to seek ways to improve them, from both the bird's and producer's viewpoint[46,54,55].

One design change made to the standard cage, 30 cm wide and 46 cm deep, was to reverse it so the birds had a longer access to the trough. This was termed a reverse cage. In one study[42], it was claimed that birds in reverse cages laid 10% more eggs which were 3.8% heavier on 2.7% less feed. There were also fewer cracked eggs and less feather loss. It appears that in the reverse cage most bird movement is forward and back rather than across the front of the cage and as a result more time is spent eating and less aggression occurs. As feeding is a social activity in hens, multi-bird cages should allow adequate trough space so that all birds feed at once. However, the full implications of the reverse cage are not clear.

Other experimental cages have been designed such as the 'get away' cage (Bareham), the Elson cage and the Brantas cage (see fig. 46).

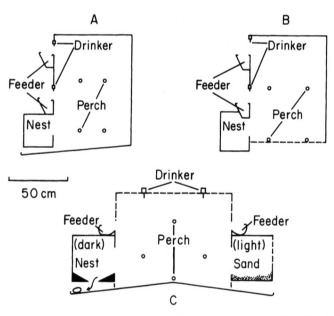

Figure 46 Three cage designs to meet the behavioural needs of laying hens. A, Brantas and Corstiaensen; B, Elson; C. Brantas (after Brantas, C.G., PRC, The Netherlands)

The Brantas cage allows a scratch box as well as a nest box. Despite the intention to improve the cage, behaviour studies have still identified some problems. Eggs being laid and broken in the scratch box was a problem. However, creeping under other hens, a major problem in cages, was reduced and birds used the roost rails in the top of the cage on which they wing stretch. Of the two troughs available 40% was taken from the top one. The greater feeding space available meant less feather pecking.

A major disadvantage of these more complex cages, apart from the greater capital cost, is that inspection of the birds is still difficult.

Solid sides in cages have reduced the amount of feather pecking of other birds through the sides and back of the cage. Feather wear on the throat on the other hand is caused by poor trough design, sharp edges and incorrect wire placement on the cage front. Horizontal bars are better than vertical bars as they allow greater food access with less pecking. Correct trough depth is also important to prevent neck blisters. The use of sand paper on the trough lip next to the bird ensures that claws are worn down and feet shape is improved.

Injuries (up to 3½%) can occur to birds such as getting their necks, wings, legs and combs caught between the bars, depending on cage design and manufacture. Manufacturers should be made more aware of these design faults.

Research highlights the need for everyone concerned in poultry production, including the manufacturers of equipment, to know the difference between density and crowding, remembering that the individual bird is most concerned about the space it carries around its head. Density is the number of birds/unit area or unit space but crowding is a product of density, contact, communication and activity. Crowding contains some psychological factors as well as physical ones.

This is illustrated by a study[1] where pecking and threats were counted in birds grouped at 4, 8, 14 and 28/cage. This provided a stocking density of 412, 824, 1442 and 2884 cm^2/bird. The results showed that the greatest number of pecks was at 824 cm^2/bird and it fell off both below and above this level. The social status of any bird in the group did not give it any advantage in feeding, activity, preening or resting. Docile hens may not necessarily be most productive.

General

Daily rhythms
The hen's day in outdoor free-living conditions starts with the rooster crowing, the descent from vertical roosting and the first feeding cycle. Egg laying takes place mid-morning and then there

is a rest period around mid-day. Feeding takes place mid-afternoon and roosting again at dusk. These behaviour rhythms are diurnal based on a 24-hour cycle but the egg-laying rhythm has a 26-hour cycle known as an 'ahemeral' cycle. As the hen treats oncoming light as dawn and darkness as night, alteration of the lighting pattern is probably the greatest tool the poultry producer uses currently to modify production. More research is needed in this whole area of diurnal rhythms.

Feeding rhythms

In caged birds working in 16 hours of daylight, three main periods of eating have been recorded. The two main ones are at the start and end of the light period and the third one was less fixed. Most is eaten at the final eating period. Broilers have been shown to eat more in a shorter time than non-meat type birds. They have longer resting times and hence convert feed to liveweight more efficiently.

When tests of feeding rhythms were made of birds in continuous light, they also engaged in regular rhythmic bouts if they could detect any changes like temperature fluctuations or feeding times to allow them to be formed. Egg-laying rhythm also helped to establish feeding rhythms.

Social feeding is important in birds and results in greater feed intake/bird. Group-fed birds are calmer and more relaxed about eating. This extra intake may be beneficial if all birds are producing, but if the habit encourages low producers to eat more, then it will be most unprofitable.

Scratching is also part of the eating ritual in free-range birds and a hungry hen scratches more than a satiated one. Running with food in the beak (food running) is seen in younger chicks but it also occurs in hens trying to get away from competitors. This habit seems to aid the breaking up of the feed while held in the beak.

Drinking rhythms

Water has been described[32] as the forgotten nutrient to the hen especially as the egg is 74% water. Egg number and size are affected within 48 hours of water being cut off. Drinking rhythm is closely related to the feeding rhythm. Neither food nor water is taken at night and intake of both peak in the final feeding of the day. Within a temperature range of 18–24°C, intakes of both are regular and stable. However, in heat waves, water intake may increase 50% and the hot afternoon is when most water is taken. As pullets grow they change their major water intake from morning to afternoon. A laying hen requires about 170 ml water/day and maximum water intake is related more to the body recovery after laying rather than to ovulation.

Water management is important when pullets are shifted to new laying quarters. Some of their previously familiar waterers should be put in their new area (this may not be practical) and most birds will copy the few that start drinking. The waterers should be started before the lights go on in the morning then the birds will drink before they eat. Air locks may occur in some systems that are stopped and re-started.

Other behavioural problems
The following are some of the problem areas of poultry production where behaviour is involved.

Ventilation
Ventilation is the key to the removal of environmental factors such as high ammonia and carbon dioxide levels, atmospheric dust, incorrect humidity and extremes of temperature. Details of ventilation needs for various classes of bird are given in the specialist texts. (See references 8, 49, 50.)

Summer heat
In very hot environments such as the tropics, special care is needed in house siting and construction to avoid excesses of the sun's radiation[53].

The birds' entire behaviour rhythm may have to be changed so that the period of maximum eating is done in the cooler part of the day. This can be done easily by changing the light–dark pattern. Birds may dig into litter which is too deep, causing further distress from the fermentation heat. As hens do not sweat, heat stress is obvious from the birds' panting, lower feed intake, increased water intake and reduced production. Lethal environmental temperatures are about 46–47°C depending on humidity. Above 39°C, severe stress is clearly shown by birds. Eggs laid during these high temperatures will have reduced hatchability.

To prevent heat stress try to do the following:

- Provide plenty of cool water. Keep pipes off the sunny walls of houses.
- Insulate the house.

- Open the doors.
- Build the house East–West rather than North-South.
- Use reflective and low-emissivity paints on inside and outside surfaces.
- Paint the roof white.
- Plant shade trees around the house and reduce the bare ground by planting grass.
- Spray water on the roof.
- Ventilate to the maximum of the fans in late afternoon.
- Clean the ventilators and open them wide.
- Put air inlets well above the ground level.
- Feed in the cooler part of the day.
- Provide a misting system within the house.
- Do not build houses too close together.

Vent pecking

In adult birds this is not related to feather pecking, occurs mainly among 18-week-old pullets about to lay, and may be associated with hormonal changes at that time. It may start when hens see others laying. There are no practical recommendations to prevent it. Some change in the environment such as dimming the lights may be tried.

Use of polypeepers

These are plastic spectacle-like clips which fix on the bird's upper beak and prevent it from seeing straight ahead. It avoids the mutilation of beak trimming to stop feather pecking and cannibalism. Although Brambell[4] advised against their use, studies in recent years mainly in Australia have shown them to be very effective in reducing pecking and peck damage, and subordinate birds had freer access to feed. Studies showed that it takes a while for the birds to adjust but after about a month of wearing them, they produced more eggs for less feed and were much quieter than before. Some producers have stopped using them as they come off and are too costly, and it is not appropriate to their management system.

Polydipsia

This is excessive water drinking by birds. It occurs in caged birds which through boredom, play excessively with drinking nipples. This leads to regurgitation of food and water attracting the attention of other birds which then often eat the mixture. This practice causes food wastage and should be discouraged by frequently disturbing the food in the trough to make it visually interesting. This can be done mechanically by a plate which ploughs through the food when pushed along.

Fowl hysteria

Hens may suddenly start flying about, squawking and trying to hide, and the episode may last for a few seconds or even minutes. This nervousness or 'hysteria' is described in reference 31. In a cage system it is confined to one or more cages while the next-door hens are not upset. In deep-litter houses and on wire floors, it can move in waves across the shed, and the result is often scratched or injured birds, pain and suffering, reduced appetite and lowered production.

It may be sparked off by handling, sudden unusual noises, bright lights reflected from car windows, untrained people entering the house, or it may appear to be spontaneous, with some strains being more prone than others. It is not usually seen in houses before 140 days of age, or in community cages until about 45 weeks of age. The only suggestion made so far for prevention is to change the design of the house in which it occurs and purchase another strain of bird. It has been suggested[31] that it is similar to the streaming observed[43] in feral chickens when the brood groups begin to break up at 18 weeks of age. Checks should always be made for vermin or predators in the poultry shed as these are usually guaranteed to cause panic. If it occurs, a complete check of all management variables should be undertaken.

Egg eating

Once birds start this habit, identifying the culprits can be very time-wasting and they can soon spread the habit to others. Indeed, experienced egg eaters will eat their own eggs and wait around for other birds to lay and eat their eggs too. It is often suggested that the habit arises because of some nutritional deficiency or shortage of grit to help digestion. The usual way hens start is by finding a broken egg and their natural curiosity pecking starts them eating.

Suggestions to solve the problem are:

- Try to identify the culprit by yolk on the beak and attempts to clean the beak.
- Cull the culprits if identified.
- Trim the beaks.
- Darken the nest boxes.
- Prevent floor eggs.
- Completely remove all broken eggs.

- Prevent soft-shelled eggs.
- Fill eggs with mustard for the culprits to eat.

Welfare

Serious concern for poultry welfare in Britain resulted in the publishing of the Brambell Committee Report[4] and this was a key document in bringing about pressure for change in systems and change in attitudes. Battery cages have continued to be the centre for both concern and research in poultry.

Measuring stress

The major concern is how to measure stress. In practical terms, birds that are not coping with their conditions will show 'stress' in its widest definition by abnormal behaviour, lowered feed intake, low production, disease incidence or even death. Recently, the degree of stress an animal is suffering has been measured by stress hormones (corticosteroids) circulating in the blood. These respond very rapidly to the stressor, they are present in blood collected from anywhere in the body and can be measured by established techniques. Capturing and handling the bird for blood sampling will also raise the levels, as shown in fig. 47.

Studies have shown that after mixing strange males which caused stress and high hormone level, although behaviour appeared normal soon after mixing, the corticosterone level peaked at 7 days after mixing, showing that stress was much longer-lasting than behaviour observations would show. The use of a product of the thyroid gland (T_3) may even be a more sensitive indicator of stress[59]. The weight

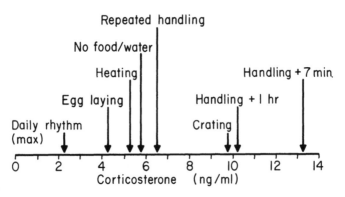

Figure 47 The levels of stress hormone (corticosterone) after different activities or treatments (after Beuving, G. (1980). In *The Laying Hen and its Environment*, ed. R. Moss, The Hague: Martinus Nijhoff Publishers)

of adrenal glands is not an effective measure, although it has been commonly used.

Increased heart rate is a well-recognised measure of stress. Sleeping birds have a rate of about 230 beats/minute which during the day ranges from 280–320. Panic can raise the level to 450 or above and during this time, blood is shifted away from the egg tract resulting in a pause in egg laying.

Selection for stress tolerance
There are now enough examples of lines of birds that have been selected to live and perform in stressful environments to show that this has real potential and should be a major technique in producing future birds.

Asking the hen
Because of the dangers of anthropomorphic interpretation of what hens need, there is now a strong interest in research which is concentrating on developing techniques of giving the hen a choice and measuring its discrimination. This research offering choices has determined the following facts:

- Hens have a slight preference for triangular entrances to nest boxes.
- Hens prefer fine netting with more points of support than the recommendations of the Brambell report.
- Hens prefer to associate with hens they know.
- Hens go to a familiar hen rather than an empty cage, and go to an empty cage rather than a strange hen.

'T' mazes have been used[12–16] to show that three out of ten hens chose a battery cage with food rather than an outside run without food, and in general hens reared in a battery cage chose to go into one rather than outside to a free, open run. A large cage with grass on the floor was preferred by hens not reared in cages.

The Skinner box is also used to get hens to make choices by pecking and from this work the preferred temperatures of hens have been worked out. The same equipment can be used to ask birds to discriminate so that their preferences for noise levels, smells and taste can be accurately assessed.

Practical welfare points
As hens will be kept intensively in future for economic reasons

managers must incorporate into their systems some practical methods to meet the animals' welfare needs. From what is currently known about the behaviour of the hen, the following points should be considered:

- Ventilation and environment control must meet the bird's needs.
- Feeding should meet the nutritional needs of the bird.
- Good quality water is necessary with two nipples/cage.
- Diets should be compounded to provide visual interest for the birds.
- Troughs should be the correct depth.
- There should be enough trough space to allow communal eating.
- Cages should be easily cleaned.
- Better cages must be designed especially to avoid injury to birds.
- Solid sides in cages reduce feather pecking.
- Engineering standards of the cage must be high – no sharp edges or rough galvanising.
- Cages must allow easy inspection.
- Cage floors must provide firm support for the bird.
- Beak trimming if considered necessary should be done with equipment that is in top-class condition, and by experienced operators. They should inspect the results of their work later to appreciate the need for high proficiency.
- Newly hatched chicks need special care during transport and in helping them to find food and water.
- TGC or tender, gentle care. Hens of all ages respond to this. Systems which ignore it are bound to fail, regardless of size.

References

1. Al-Rawi, B. and Craig, J.V. (1975) 'Agonistic behaviour of caged chickens related to group area and area per bird.' *Appl. Anim. Ethol.* 2, 69–80.
2. Bakhuis, W.L. (1974) 'Observations on hatching movements in the chick (*Gallus domesticus*).' *J. Comp. Physiol. Psychol.* 87, 997–1003.
3. Bell, D. (1980) 'Poultry management for today's industry.' *N.Z. Poult. Wld.* 44(4), 31–2.
4. Brambell, F.W.R. (1965) *Report of the technical Committee to inquire into the Welfare of Animals kept under intensive Livestock husbandry Systems.* London: HMSO, Cmd, Paper No. 2836, 85pp.
5. Brantas, G.C. (1977) 'Some observations about the newest type of get-away cage for laying hens.' *Proc. 28th Ann. Meet. Europ. Assoc. Anim. Prod. Brussels*, pp. 1–4.
6. Brantas, G.C. (1980) 'The pre-laying behaviour of laying hens in cages with

and without laying nests.' In *The laying Hen and its Environment*, ed. R. Moss, The Hague: Martinus Nijhoff Publishers, pp. 227–37.

7. Candland, D.K. (1969) 'Discriminability of facial regions used by the domestic chicken in maintaining the social dominance order.' *J. Comp. Physiol. Psychol.* **69**, 281–5.

8. Charles, D.R. (1981) 'Environment for poultry.' *Vet. Rec.* **106**, 307–9.

9. Clark, J.A. (1981) *Environmental Aspects of Housing for Animal Production*, London: Butterworths, p. 511.

10. Craig, J.V. (1980) *Domestic Animal Behaviour: Causes and Implications for Animal Management*. New Jersey: Prentice-Hall Inc., pp. 1–364.

11. Craig, J.V. and Bhagwat, A.L. (1974) 'Agonistic and mating behaviour of adult chickens.' *Appl. Anim. Ethol.* **1**, 57–65.

12. Dawkins, M. (1975) 'Towards an objective method of assessing welfare in domestic fowl.' *Appl. Anim. Ethol.* **2**, 245–54.

13. Dawkins, M. (1977) 'Do hens suffer in battery cages? Environmental preferences and welfare.' *Anim. Behav.* **25**, 1034–46.

14. Dawkins, M. (1978) 'Welfare and the structure of a battery cage: size and cage floor preferences in domestic hens.' *Brit. vet. J.* **134**, 469–75.

15. Dawkins, M. (1980) 'Environmental preference studies in the hen.' *Anim. Regul. Stud.* **3**, 57–63.

16. Dawkins, M.S. (1982) 'Elusive concept of preferred group size in domestic hens.' *Appl. Anim. Ethol.* **8**, 365–75.

17. Dawson, J.S. and Siegel, P.B. (1967) 'Behaviour patterns of chickens to ten weeks of age.' *Poult. Sci.* **46**, 615–22.

18. Duncan, I.J.H. (1980) 'The ethogram of the domesticated hen.' In *The laying Hen and its Environment*, ed. R. Moss, The Hague, Martinus Nijhoff Publishers, pp. 5–18.

19. Elson, H.A. (1979) 'Design of equipment for feeding the bird.' In *Food Intake Regulation in Poultry*, eds. K.N. Boorman and B.N. Freeman, Edinburgh: Brit. Poult. Sci. Ltd, pp. 431–44.

20. Faure, J.M. and Bryan-Jones, R. (1982) 'Effects of sex, strain and type of perch on perching behaviour in the domestic fowl.' *Appl. Anim. Ethol.* **8**, 281–93.

21. Faure, J.M. and Bryan-Jones, R. (1982) 'Effects of age, access and time of day on perching behaviour in the domestic fowl.' *Appl. Anim. Ethol.* **8**, 357–64.

22. Fischer, G.J. (1975) 'The behaviour of chickens.' In *The Behaviour of Domestic Animals*, 3rd edn, ed. E.S.E. Hafez, Baltimore: The Williams and Wilkins Co., pp. 454–89.

23. Folsch, D.W. (1980) 'Essential behavioural needs.' In *The laying Hen and its Environment*, ed. R. Moss, The Hague: Martinus Nijhoff Publishers, pp. 121–47.

24. Gross, W.B. (1981) 'Some effects of adapting chickens to their handlers.' *J. Amer. vet. Med. Assoc.* **179**, 270 (abstr).

25. Gross, W.B. and Siegel, P.B. (1979) 'Adaptation of chickens to their handlers and experimental results.' *Avian Dis.* **23**, 708–14.

26. Guhl, A.M. (1953) 'Social behaviour of the domestic fowl.' *Tech Bull. Agric. Expt Stat.* Kansas State Coll. Manhattan, No. 73. pp. 1–48.

27. Guhl, A.M. (1968) 'Social inertia and social stability in chickens.' *Anim. Behav.* **16**, 219–32.

28. Guhl, A.M. and Ortman, L.L. (1953) 'Visual patterns in the recognition of individuals among chickens.' *Condor* **55**, 287–98.

29. Hale, E.B. (1948) 'Observations on the social behaviour of hens following debeaking.' *Poult. Sci.* **27**, 591–2.

30. Hann, C.M. (1980) 'Some system definitions and characteristics.' In *The laying Hen and its Environment*. ed. R. Moss, The Hague: Martinus Nijhoff Publishers, pp. 239–58.

31. Hansen, R.S. (1976) 'Nervousness and hysteria of mature female chickens.' *Poult. Sci.* **55**, 531–43.

32. Hearn, P.J. and Hill, J.A. (1978) 'The role of water in commercial egg production.' *A.D.A.S. Quart. Rev.* **28**, 35–50.

33. Hembree, D.J., Adams, A.W. and Craig, J.V. (1980) 'Effects of force-moulting by conventional and experimental light restriction methods on performance and agonistic behaviour of hens.' *Poult. Sci.* **59**, 215–23.

34. Hou, S.M., Boone, M.A. and Long, J.T. (1973) 'An electrophysiological study on the hearing and vocalisation in *Gallus domesticus*.' *Poult. Sci.* **52**, 159–64.

35. Hughes, B.O. (1975) 'The concept of an optimum stocking density and its selection for egg production.' In *Economic Factors affecting Egg Production*, eds. B.M. Freeman and K.N. Boorman, Edinburgh: Brit. Poult. Sci. Ltd. pp. 271–98.

36. Hughes, B.O. (1980) 'Behaviour of the hen in different environments.' *Anim. Regul. Stud.* **3**, 65–71.

37. Hutchinson, J.C.D. and Taylor, W.W. (1962) 'Mechanics of pecking grain.' *Proc. 12th Wld Poult. Congress.* 112–16.

38. Jaksch, W. (1981) 'Euthanasia of day-old male chicks in the poultry industry.' *Int. J. Stud. Anim. Prob.* **2**(4), 203–13.

39. Kilgour, R. (1975) 'Aspects of animal behaviour in relation to intensive production.' *Proc. 3rd Comb. Conf. Aust. Chicken meat Fed*, Aust Stock Food Manuf. Assoc. pp. 96–106.

40. Kite, V.G., Cumming, R.B. and Wodzicka-Tomaszewska, M. (1980) 'Nesting behaviour of hens in relation to the problem of floor eggs.' In *Behaviour in relation to Reproduction, Management and Welfare of Farm Animals*. Rev. Rural. Sci., IV Univ New England, N.S.W.. pp. 93–6.

41. Laughlin, K.F. (1981) 'The incubation of eggs.' In *Environmental Aspects of Housing for Animal Production*, ed. J.A. Clark, London: Butterworths, pp. 103–6.

42. Lee, D.J.W., Bolton, W. and Dewar, W.A. (1978) 'Effects of battery cage shape and dietary energy regulation on the performance of laying hens offered diets containing dried poultry manure.' *Brit. Poult. Sci.* **19**, 607–22.

43. McBride, G., Parer, I.P. and Foenander, F. (1969) 'The social organization and behaviour of the feral domestic fowl.' *Anim. Behav.*, Monogr, **2**, 127–81.

44. Morris, T.R. (1981) 'The influence of photoperiod on reproduction in farm animals.' In *Environmental Aspects of Housing for Animal Production*, ed. J.A. Clark, London: Butterworths, pp. 85–102.

45. Murphy, L.B. (1980) 'Drinking facilities for brooding broilers.' In *Behaviour in relation to Reproduction, Management and Welfare of Farm Animals*. Rev. Rural. Sci., IV Univ New England, N.S.W., pp. 147–49.

46. Preston, A. (1980) 'Evaluation of animal environments using behaviour.' In *Behaviour in Relation to Reproduction, Management and the Welfare of*

Farm Animals. Rev. Rural. Sci., IV Univ New England, N.S.W., pp. 151–2.

47. Ross, P.A., Hurnik, J.F. and Morrison, W.D. (1981) 'Effect of controlled drinking time on feeding behaviour and growth of young broiler breeder females.' Poult. Sci. 60, 2176–81.

48. Rumsey, R. (1978) 'Chrome tape attracts pullets to nests.' N.Z. Poult. Wld. 42(2), 21.

49. Sainsbury, D.W.B. (1980) 'Poultry production and welfare.' Anim. Regul. Stud. 3, 43–9.

50. Sainsbury, D.W.B. and Sainsbury, P. (1979) Livestock Health and Housing, 2nd edn., London: Bailliere Tindall, 388pp.

51. Scott, W.N. (1978) 'The slaughter of poultry for human consumption.' Anim. Regul. Stud. 1, 227–34.

52. Siegel, P.B. (1978) 'The interfacing of breeding and behaviour in poultry.' In Hohenheimer Arbeiten, Eugen Ulmer, Stuttgart, 93, 9–24.

53. Smith, W.K. (1981) 'Poultry housing problems in the tropics and subtropics.' In Environmental Aspects of Housing for Animal Production, ed. J.A. Clark, London: Butterworths, pp. 235–58.

54. Tauson, R. (1980) 'Cages: how could they be improved?' in The laying Hen and its Environment, ed. R. Moss, The Hague: Martinus Nijhoff Publishers, pp. 269–304.

55. Tauson, R. (1981) 'Need for improvement in construction of cages.' Proc. 1st Europ. Sympos. Poult. Welfare. June. pp. 61–74.

56. Van Kampen, M. (1981) 'Thermal influences on poultry.' In Environmental Aspects of Housing for Animal Production, ed. J.A. Clark, London: Butterworths, pp. 131–47.

57. Vestergaard, K. (1981) 'Dust-bathing behaviour in the domestic hen.' Appl. Anim. Ethol. 7, 386–7 (abstr).

58. Williams, N.S. and Strungis, J.C. (1979) 'The development of grooming behaviour in the domestic chicken (Gallus gallus domesticus)'. Poult. Sci. 58, 469–72.

59. Wodzicka-Tomaszewska, M., Cumming, R.B., Ambuhl, S., Snell, M.L. and Stelmasiak, T. (1980) 'Some measurements of stress in intensively-housed poultry.' In Behaviour in relation to Reproduction, Management and Welfare of Farm Animals, Rev. Rural. Sci., IV Univ New England, pp. 187–9.

60. Wood-Gush, D.G.M. (1971) 'The pre-laying behaviour of the domestic hen.' Span. 14(2), 105–7.

61. Wood-Gush, D.G.M. (1971) The behaviour of the domestic fowl, London: Heinemann Educational Books, 147pp.

62. Wood-Gush, D.G.M. (1975) 'Nest construction by the domestic hen: some comparative and physiological considerations.' In Neural and Endocrine Aspects of Behaviour in Birds, eds P. Wright, P.G. Caryl and D.M. Vowles. Amsterdam: Elsevier Scient. Publ. Co, pp. 35–49.

Further reading

Curtis, S.E. (1981) Environmental Management in Animal Agriculture. Illinois: Animal Environment Services, p. 385.

Wood-Gush, D.G.M. and Duncan, I.J.H. (1976) 'Some behavioural observations on domestic fowl in the wild.' Appl. Anim. Ethol. 2, 255–60.

9 Dog

The dog was one of the first species of animal kept in close relationship with man. As well as being used for protection, hunting, sport and companionship it also provided food, clothing and traction. The dog still provides many of these functions today.

Behavioural definition

The dog is a pack animal that keeps close social contact with other dogs through barking and howling, smell, touch and to a lesser extent sight. The capture of large prey requires co-operative group hunting skills. The bitch gives birth to relatively undeveloped pups in a den and spends much of her time with them for the first few days. After this however she returns to the hunting pack to assist, leaving the pups to themselves for long periods. The pups at birth are blind and deaf and operate largely on touch. After the eyes and ears open, the pups start to interact and the sensitive period for socialisation is from weeks 4 to 10. As a pack animal the dog is motivated by reward and reprimand, showing its social rank by dominance or submissive body posture and associated behaviour. After puberty dogs exhibit territorial behaviour and may attack intruders, though to some extent these responses are less developed than in wild canines. Territorial marking, sniffing and urinating on posts or other objects are normal adult responses.

Mating results in the copulatory tie and bitches once mated often show a strong preference for mating with the same male again. They are seasonal breeders with less shift to year-round breeding than is shown by many other domestic species. The dog has a wide dietary preference and its reliance on smell for food location suits its lifestyle as a scavenger.

Birth

Behaviour before birth

As birth approaches, most bitches become restless and their appetite is reduced. About 12–24 hours before birth a bitch will try to make a bed in a place other than her normal kennel. If shut in her normal kennel she paws the floor and sides, but given bedding (straw, paper or rags), a bitch will chew it up and re-arrange it to make a nest.

A working dog will still want to work, even at this late stage of pregnancy, but she should be watched carefully and protected from being injured by stock. She may suddenly disappear, returning home to make a bed. She may even start having her pups while at work. If this happens, she should be returned to her whelping quarters immediately.

Behaviour during and after birth

The length of labour can vary considerably from 2 to 3 hours to 12 hours or more, depending on the size of the litter and the amount of disturbance experienced by the bitch. Studies[2] have recorded that in normal births the bitch lies on her side and periods of rapid respirations (100–175/minute) alternate with periods of slower, deeper respirations (40–60/minute). Normal resting rates are 16–20/minute. Eyes may be half-closed and often turned upward. The bitch periodically sniffs and licks her genitalia and during contractions, twitching of the hind legs and slight to moderate shivering accompanies uterine contractions. The pups are expelled by forceful straining of the abdominal and thoracic muscle. This process continues with the birth of each pup, and ends when separate afterbirths are expelled. The bitch usually has a brief rest between each pup's birth. The dam may move about between the birth of pups to drink or urinate, but she returns to her laying position. Most bitches prefer to whelp when lying in their beds, although sometimes they may rise and move about while straining.

As soon as the pup is delivered, the bitch chews at and eats the birth sac, biting through the umbilical cord in the process. The pups may stick in the birth canal until pushed out by the following pup during contractions. Dead pups are licked and attended like live ones until they become cold and are then ignored and may be pushed out of the way. Normally, they are not eaten.

The bitch spends a lot of her time keeping the nest area clean and sanitary. She does this frequently, licking her pups and ingesting their faeces and urine, which are voided when she licks their lower abdomen and perianal areas.

During whelping the bitch rarely barks or whines, and she gives her full attention to nursing after the last pup has been born and she has rested. Care should be taken by strangers inspecting her at this stage and handling the pups. They should be discouraged from visiting bitches with their first litters for a few days after birth. Their presence, and the distress call of a pup, may make the bitch aggressive. The dog's owner should always accompany strangers as the bitch may bite them or savage her pups.

Disturbance during whelping

Studies[2] showed that disturbance of a bitch during birth caused interruptions in the birth process and long delays between pups.

In normal whelpings, the average observed time between births was 30 minutes but disturbance delayed this up to several hours. In practice, this means that the noises and disturbance of the working day will normally delay the birth of most of the pups until evening. These studies also showed that bitches varied in the amount of aggression and stress caused by disturbance.

Behaviour of pups after birth

Pups are born blind and deaf and unable to stand but are quite active soon after birth, having been stimulated by their dam's grooming, sometimes to a point of apparent exhaustion.

A new-born pup starts to crawl short distances around the nest with frequent rests, when it makes exploratory side-to-side head movements, searching for the teat. The bitch may encourage and direct it to the teat by licking it. If the bitch is still giving birth to other pups, then she does not concentrate on the pups that are teat-seeking.

If pups contact cold surfaces, they usually retreat until they find the warm body of the bitch, where they start burrowing and head-moving, trying to locate a nipple.

Their first vocalisations are squeaks and muffled grunts to express discomfort, pain or hunger. Barking has been observed as early as 18 days in pups. Playing freely with litter mates seems to encourage pups to bark. The approach of strangers when pups are spread out playing will also stimulate them to bark as an alarm call.

Pups like to feel warm, soft surfaces rather than cold, hard ones and holding a pup close to the owner's body will usually calm it. This 'contact comfort' has been shown[8] to be important in dogs. Any pups that wander from the nest and get lost vocalise loudly and the bitch will carry them back to the nest holding them in her mouth. Within the confines of the nest she can attract them back to her body by licking them.

When the bitch lies down among a dispersed litter of pups there is a risk that she may lie on or smother some of them. She usually ignores short cries from the pups but investigates persistent cries and hence finds pups that are being crushed. Teat-seeking is signalled by a low grunting noise. Fitting a low rail could help to protect pups from being crushed by inexperienced bitches.

Rearing

Many behaviour studies have highlighted how, just before 4 weeks old, the puppy begins to interact with its litter mates and its mother. In the wild the mother would often be absent at this stage. At this stage, the pup can now see and hear, and play. Tail wagging, barking and other characteristics of the adult dog are also expressed. By 4 weeks old they stand and move like adults. Weaning is at 6 weeks.

Socialising the pup

From 4 weeks old to about 12 weeks the pup establishes social relationships with other dogs and people. If the pup is only socialised to people it will be difficult to breed. If it is only socialised to dogs it will be anti-social towards people and be difficult to train. Hence at this stage, the pup should be exposed to all the variables it will meet later in life, including other dogs, animals, vehicles, different people, children, noise and so on. The pup cannot be given too much experience and reassurance by its owner at this stage. The belief of some working dog trainers that pups can be spoiled by children is not supported. It is very important that a working dog is socialised with children to avoid biting accidents later in its life.

Pups up to 5 weeks old readily approach strangers but after this age they avoid them. Studies[4,5] showed that this avoidance phenomenon reaches a peak at about 8 weeks, with some variation among breeds and strains. It seems to be nature's way of protecting the young animals from predators by an 'anxiety period' of avoidance behaviour. So what a pup has not experienced before 12 weeks old will be avoided and feared subsequently.

It was also shown that learning is unstable in young pups 5–6 weeks old, but more stable at 8–9 weeks old. This 8–9 week age is a very sensitive time for pups. Pups which were punished at this age never forgave the experimenter, whereas pups punished at 12–13 weeks did. These older pups were emotionally attached to the experimenter, irrespective of punishment.

Thus 4–12 weeks is the most critical and formative period of a pup's life. The best period to take a pup as a pet is between 6–10

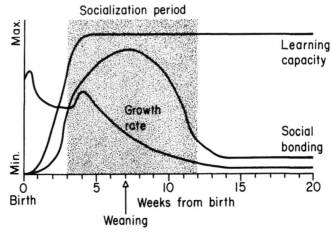

Figure 48 Aspects of a pup's development from birth to 20 weeks old (after Scott, J.P., Stewart, J.M. and de Ghett, V.J. (1974), *Develop. Psychobiol.* 7, 509–13)

weeks old, so that it has both social experience with its own kind but is still able to develop an emotional bond with humans. Various aspects and times of the pup's development are shown in fig. 48.

The main point to remember during rearing is that a dog is a pack animal, and most pack animals are co-operative and fit into a clearly defined hierarchy, either with other dogs or a mixture of dogs and humans.

Once the bitch discourages pups to nurse, the focus is shifted from her teats to her muzzle. Pups lick the muzzle while assuming the crouching body position of a low-ranking pack member (fig. 49). In wolves, for example, and the wild dog, this induces regurgitation.

Figure 49 Pups approach the muzzle of a dog silhouette painted on a wall (after Fox and Weisman (1970). *Develop. Psychobiol.* 2, 277)

of partially digested food by the adult. It is not such a common feature of a domesticated working dog, though the basic approach is still nose-to-nose.

A rearing programme

Using the behavioural responses of the dog, a rearing programme has been described[3,17] which uses non-verbal physical handling exercises to develop a dominant–subordinate relationship with the pup and establish authority through actions rather than words. These exercises simulate some of the handling that is done by the bitch and other pack members. They are:

Exercise 1: Elevation

Lift the pup up to your eye level by holding it behind its front legs while supporting its body for periods of 30–90 seconds. Hold it away from you and look directly into its eyes. If it struggles, shake it and raise your voice. Praise the dog softly when it settles. Repeat the exercise for varying periods in different places and in the presence of other people, for a few weeks.

Exercise 2: Inversion

Hold the pup by the scruff of its neck with one hand and turn it upside down. Hold it below its rump with the other hand. Hold it away from you, and if it struggles, shake it. Reassure it when it lies quietly.

Exercise 3: Straddling

Straddle the dog from behind as if to ride on its back. Lock your hands under the dog's chest just behind its front legs and lift it off the floor. It it struggles, shake it and raise the voice. When it settles, praise it in a quiet voice. Repeat at different times in different places.

Exercise 4: Prone

Hold the pup on its side on the floor. If it struggles, shake it. If it lies quietly, praise it. Handle all parts of its anatomy, including its mouth, putting your fingers inside.

Carrying out these exercises ensures that the pup will assume a subordinate role to its master and other people. The handling should be firm but gentle. Remember that asserting dominance does not require an aggressive approach with physical beating. The leader–

follower bond is not based on fear, and in dog packs the low-dominance animals seek out the top-dominant animal for body contact.

Discipline of pups for misdemeanours must be done immediately they are caught in the act. To correct a pup, a quick shake with raised voice is sufficient. It should then be quietly praised to provide positive reinforcement to re-establish a good relationship.

Correct human—dog relationsips can be built up by the following practical steps[17]:

- Feed the pup yourself so the pup associates you with control of its food supply.
- Occasionally interrupt its feeding for a few seconds, praise the pup and replace the food. If it resists, use the shake-praise procedure.
- Never deliberately call the dog to you to administer discipline.
- Avoid being separated from the pup for long periods of time during its early development. Take it with you whenever possible, but take care it does not get injured or too frightened.
- Introduce it carefully to traumatic situations such as noise or action. Keep it in a safe position near you to reassure it.
- Do not change the rules — be consistent with the pup and praise it regularly. If it shows dominance or develops bad habits, use the shake-praise procedure. Examples of faults are nipping or snapping, growling, barking when not wanted, mounting people's legs, etc.
- Good non-fearful signs of a subordinate dog are a vigorous welcome, ears held back with head and body slightly lowered, tail held down but wagging, mouth open and lips drawn back in a grin and licking of the owner's hands and face.

Reproduction

The male

Most working dogs are reared in contact with other dogs and during play, develop the techniques required for mounting and mating. A dog reaches puberty at about 6—9 months old.

Dogs show great individual variation in their mounting and mating behaviour. If the dog is brought to the bitch, he may need more time to settle into her environment before mating than if the bitch was taken to his territory. During this time of adjustment keep outside distractions to a minimum.

Approaching a female in oestrus, the male smells and licks the genital area of the bitch. She responds by 'presenting' or showing the receptive posture. Some males may approach the head of the bitch and adopt the play posture before mounting. The female may respond to this by running a short distance, then presenting again.

The mating act

If the bitch is in standing oestrus, the dog will quickly mount her several times. Then he will start pelvic thrusting and gain intromission by trial and error. He may attempt several pelvic-thrusting mounts before complete entry is achieved.

The dog then pulls its front legs back and appears to put its tail between its legs. While thrusting vigorously, the dog steps alternately from one back leg to the other, and it is during this activity that ejaculation occurs[6]. During this reaction, which lasts 15–30 seconds, the penis engorges within the vagina to lock the male and female together.

The bitch usually stands rigidly during the most intense ejaculatory reaction but may cause the male to fall off by twisting and turning. Some males may dismount and by stepping over their penis, end up in a tail-to-tail lock position which may last for 10 to 30 minutes. During this lock position there is a second ejaculatory reaction when most of the prostatic fluid is expelled[6]. Responses during mating are shown in fig. 50.

The female

The bitch has two heats a year, and urinates to mark her territory

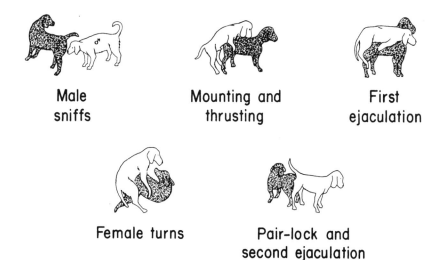

Male
sniffs

Mounting and
thrusting

First
ejaculation

Female turns

Pair-lock and
second ejaculation

Figure 50 The mating sequence in dogs[6]

when in heat. She has a lengthy period of pro-oestrus (when she is coming into heat) and the standing heat lasts about 5–12 days.

If not pregnant after mating, a pseudopregnancy of 30–90 days may develop, after which the bitch will not show oestrus for 3–4 months. During pseudopregnancy, the uterus expands and mammary development and milk production can occur as in normal pregnancy. Near the time of expected birth the bitch may make a nest and will hide and mother objects and lactate for up to 60 days.

Signs of oestrus

In working dogs the first signs that a bitch is coming into oestrus is the interest shown in her by male dogs. They will smell her genital area and places where she has urinated, showing great excitement and agitation if more than one male is present.

The vulva of the bitch will be red and swollen and viscid mucus and blood may be shown. As oestrus continues, she will show greater interest in the males and stand to be mounted. She may stand and then race around in a play-like action to tease the male who may solicit this behaviour himself.

The bitch may urinate in the presence of the male, and this is a good sign that she is receptive. She may do this by a male leg-lift posture. The bitch in heat may mount the dog with pelvic thrusts if he is slow in making courtship advances. Other females will also mount her.

The receptive posture of the bitch is characterised by a lateral curving of the hind quarters and arching the tail to one side as the male sniffs at her. When mounted the bitch stands rigidly. After the dog's penis has expanded the bitch begins to twist and turn for about 5–30 seconds, often throwing the male off. She may roll over once or twice and yelp[6].

Castration of dogs

Adult male dogs are sometimes castrated in attempts to reduce their roaming tendencies, aggression, urine-marking habits, mounting other dogs or people, as well as to prevent them breeding.

Research studies[7] showed that there was great variation between individual dogs in their response to the effects of castration. Roaming was reduced in 90% of dogs castrated at a minimum age of 8 months. The other behavioural patterns of aggression, urine-marking and mounting were reduced by one-half to two-thirds in the 42 castrate dogs observed. Age of the dog had no effect on the dog's reaction to the operation. However, it should be noted that the reduction in urine marking was in the dog's own territory (the owner's house). Scent marking is not reduced in other novelty areas by castration[6].

Although dogs will copulate for a considerable period (4 months) after castration, there is a marked drop in their response to receptive females after about 2 months. They cannot achieve intromission or a copulatory lock.

Speying of bitches

This surgical operation involves the complete removal of ovaries and uterus. As a result the bitch will not show oestrus during the rest of her life and is completely sterile. Some speyed bitches become excessively fat in later life and sluggish in temperament.

Urinating and defecating behaviour

Females from puppyhood until maturity urinate in a squatting position. Bitches with pups always urinate and defecate away from the nest. Bitches do not urinate to mark their territory to the same extent as males.

In the male, however, urination (and to a lesser extent defecation) is very much involved in scent-marking as well as evacuating the bladder and rectum. Male pups urinate standing on all fours up to about 5 weeks of age, when they start to extend their hind legs. As the pup grows older, it gradually starts to lift or cock either hind leg, twisting the body slightly to direct the urine flow onto the selected object. This also signals approaching sexual maturity and claims to territory.

Although scent-marking by urinating is a territorial behaviour pattern, the domestic dog is not able to maintain this territory as do wild dogs. It makes the territory familiar to itself and informs others of its presence.

Dogs of both sexes prefer to defecate away from their home kennel area if they can. In cities urban dogs foul streets, gardens and parks. Although the domestic dog does not have as well-developed anal scent glands as the wolf, coyote and wild dog, there is still a strong element of scent marking while defecating. Some dogs will back onto rocks, tree stumps and into thistles and bushes before defecating. Invariably, after defecating and urinating, they scratch the ground with their hind legs, throwing up earth. City dogs even do this on concrete pavements or inside on carpets.

Pups can be house-trained by restricting them to a small part of a room and putting newspapers in the corner for elimination. Once the pup has adapted to this practice, it can be given more space.

It should also be taken outside in the morning and after sleeping, when the chances of it urinating are high. Eventually the pup accepts that the whole house area is its nest or home area and will indicate by whimpering that it wishes to urinate or defecate. If this programme breaks down, it is generally due to some emotional problem such as neglect.

Semen collection from dogs for AI

Researchers[16] have listed some important criteria to assist the collection of semen from sexually inexperienced dogs. They are:

- All equipment should be clean, dry, pre-warmed and assembled.
- An isolated, quiet room is required with either non-slip flooring or a rubber mat of sufficient size for both the teaser bitch and the male.
- An anchored table leg or ring in the wall 12 inches from the floor is useful to secure the teaser bitch for collection purposes.
- Only the necessary personnel should be in the room and there should be no talking or extraneous noise or interruptions during the collection procedures.
- For inexperienced studs, an in-heat teaser bitch should be used.
- The bitch should be muzzled if there is any possibility of her biting. She can be further restrained by a handler holding her from the right side with his arm under the bitch's chest. This also prevents her from sitting down when mounted. The latter will upset an inexperienced male.
- The male dog is then led in and allowed to become familiar with the bitch. His attention should be directed to the bitch's hind quarters.
- The male is quietly but firmly encouraged to achieve an erection by manual manipulation of the penis and to ejaculate into a rubber cone attached to a graduated plastic centrifuge tube.
- If a dog shows disinterest or fright, he should be removed from the room and attempts should be made to collect him without any dogs being present.
- After a number of successful collections, the dog is encouraged to breed naturally. His first naturally mated bitch should be quiet, in standing heat and of a similar size.
- In some countries legislation requires veterinary supervision.

The dog's senses

Sight

Dogs vary greatly in the position of their eyes. In general they have up to 70° greater peripheral and 20° narrower binocular vision than man, hence they are more aware of movement at the side of their head, and they need to move their heads more in order to focus. A dog's ability to see movement is ten times as sensitive as man's. A dog receives visual cues from its environment that a man cannot see[3].

Slight movement can over-stimulate a dog to bite if it has been beaten in the past with hand or stick or kicked.

Visual discrimination of detail and silhouette is not as fine in the dog as in man, but dogs can differentiate between black and white (brightness).

Hearing

Dogs' hearing ranges from 20 000 to 50 000 cycles per second (c.p.s.), whereas human hearing peaks at about 20 000 c.p.s.. Thus silent dog whistles are sometimes used for gun dogs. A sound normally heard by man at 6 m can be heard by most dogs at 25 m.

A visual clue such as pointing an arm or a stick helps a dog to locate where the sound came from and reinforce its meaning. This 'show-and-tell' technique[3] should be exploited in training.

Studies also show that a sound stimulus is masked or reduced in effect when accompanied by touching the dog's body. Auditory stimuli like word commands must be kept simple, clear and not confuse the dog. This especially applies to the dog's name.

Dogs are very sensitive to voice tone and commands should generally be given quietly so that loud tones can be reserved for reprimands.

Dogs respond greatly to a 'happy voice'[18]. This can be used as an important tactic to divert a dog's attention from some error, and after a reprimand, to let it know that relationships have returned to normal.

Touch

As pack animals born in litters, dogs receive many stimuli through touch. Touch is important in developing owner–dog relationships. The stimulation given a dog through touch seems to be greater than through speaking to it or letting it smell you. Dogs react very quickly to touch, and touching the upper body, especially the shoulder, is an act of dominance, while touching the chest or under-side is a more submissive gesture on your part towards the dog.

Smell

The dog has a highly developed sense of smell which is used to man's advantage. Examples are locating lost people and drugs. There are also disadvantages such as dogs biting strangers who smell differently, those which leave home to find bitches in heat and find carrion to roll in. It is impossible to stop a dog smelling by punishment. It can only be taught to sniff in certain areas and not sniff in others.

Approach patterns when dogs meet and begin social contact are shown in fig. 51. Various glands, hair pattern and eye and tail movements are all involved (see fig. 52).

Barking

Barking is considered to have the primitive function of being the dog's alarm call. Dogs can produce varied barks, yelps and howls, and their significance has not been fully investigated. One researcher[15] concluded from an acoustic analysis that dogs appear to communicate by vocalisation the way humans use language as they can learn to use a variety of noises to achieve certain goals.

Training

All trainers agree that it is a waste of their time to attempt to train a dog for work if it does not have the genetic potential to learn the tasks expected of it. As this cannot be judged in a weaned group, heavy reliance has to be put on the parents' pedigree and performance. In highly selected working strains of dogs it is usually the trainer who sets the limitations and there is very wide variation in people's ability to train dogs. People with little patience, short tempers and

One dog
approaches,
starts the circling

Dominant starts
sniffing at
genital area

Both dogs
investigate

Figure 51 Responses shown by two strange dogs meeting (after Fox, M.W. (1971). *The Behaviour of Wolves, Dogs and Related Canids*, London: Jonathan Cape, p. 200.)

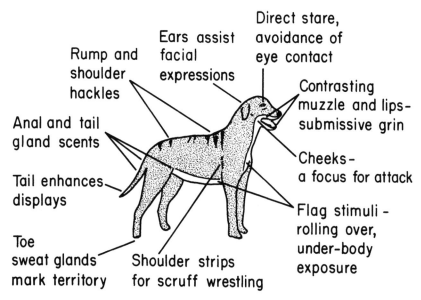

Figure 52 Parts of a dog's body used to signal social responses (after Fox (1971) – see fig. 51 caption)

who basically do not like dogs, should never start to train a young pup. The dog is very responsive to the moods of its trainer.

Selecting the pup

There are no behavioural traits in newly weaned pups that have been clearly identified as being good indicators of learning ability. There is, however, a lot of folklore. If you have a choice, trainers advise picking a bold, active pup that you like the look of, and avoiding the runt of the litter. However, there are many stories of runts turning out to be champions.

After removal from the dam at weaning (6–8 weeks old) the pup should be taken home to a warm bed, fed regularly and helped to settle into the new environment by a lot of physical rubbing and fussing. If it yaps or howls, feeding or providing warmth and comfort will generally settle it again during the first few weeks. This should stop habitual barking. The owner must be fond of the pup. This is essential in forming a good relationship.

Early training

The main points to consider during this period from weaning (6–8 weeks old) to 6 months when the dog starts to show an inclination for real work are:

- Remember the pup is a pack animal that greatly values social contact with other dogs and people. Ensure this is provided.
- Careful use of voice-tone, petting, touching and feeding are high-value rewards for dogs.
- Do not over-stimulate (over-excite) or tease the pup.
- During this early playful stage, the main aim is to try to keep the pup occupied. Do not shut it up or leave it isolated on a chain by itself. It will develop bad habits such as barking, biting and shyness for work through boredom.
- Keep the pup in a run (2 m² minimum area), or tie it up on a chain to its kennel. Having pups on a chain often encourages the owner to handle it more than when it is shut in a pen.
- Teach the pup its name. Every time you pass its run or walk to inspect it, again call its name. This becomes an alerting signal. Always use its name when feeding.
- When the pup is outside and wanders away from you, shout its name and call it with a 'here' command. Use a 'here' whistle and welcome it to you when it arrives.
- Expose the pup to a wide range of experiences in the environment but protect it from injury. The most likely causes of injury are mature farm livestock and vehicles.
- Get the pup used to wearing a light collar, and when it accepts this, tie it up for a short period (about 30 minutes) each day on a short chain with a swivel in it. Encourage and pet the animal, using its name. Ensure that it does not learn to slip the collar, as this can develop into a serious problem. Check the collar for tightness as it grows. A good test is to be able to move one finger between the dog's neck and collar.
- Get the pup used to going back to its run or kennel so that it enjoys the experience and does not have to be caught and dragged there.
- The pup may have to be taught to jump up if the kennel is above ground. Provide a step so that it does not hurt itself or fall; otherwise it will not want to go in. Make sure the pup is taught to go into the run or kennel without being fed.
- Teach the pup to lead on a check cord (about 4 m long) for the next few months, and get it to follow at heel with an appropriate word command (e.g. 'get in').
- As early as possible, teach the 'sit' by pressing down on its hindquarters at the command 'sit'. Encourage and pet it when done successfully. Some shepherds teach the 'stand' and also the 'lie down'. The latter is easily taught by stepping on the rope near

the collar to force the pup down on its belly while giving the command.

- The command 'stay there' or 'stand' can be taught while on the check cord, by leaving the dog in a set position while you move around and away from it. Do not frustrate the pup by leaving it too long. Call him with the 'come here' whistle. Keep all lessons short.
- Between these lessons there is no need to be very restrictive. The pup should be played with and fussed over but keep a careful watch for the development of bad habits like car, tractor or motorbike chasing, biting horses' heels or jumping up at people.

The use of voice tone is the key to this early training. Shouting or growling at the pup, along with a shake for misdemeanours, and a quiet and happy voice for praise are the basic tools. Patting or scratching the chest is recommended by Barbara Woodhouse[18] in preference to patting the head and rubbing the dogs' ears vigorously. This over-excites them. By this time the dog should know its name and respond to the 'come here', 'sit' and 'stand' commands. It should have demonstrated a keen interest to start work.

Trying to teach a pup basic commands by tying it to an older dog is not recommended. Dogs vary so much that little is achieved and there is a high chance that both dogs will be spoiled.

One very successful New Zealand trainer of sheep dogs, gun dogs and dogs for obedience trials (Mr Alan Lourie) teaches the basic sit, stand and stay commands with the pup on a stool about 250 mm wide, 800 mm long and 600 mm high. These basic lessons can start as early as 10 weeks and are taught 3 times a day, for 3 minutes a lesson away from the distraction of other dogs or farm events. The pup is taken from its run without fuss, placed on the stool and then fussed profusely. At the end of the lesson it is fussed on the stool then returned immediately to its pen.

This technique shortens the initial training period as the pup concentrates on its work and learns quickly. The stay command is taught very effectively as the trainer can walk away from and around the pup very easily, teaching it the discipline of staying until given further commands. The Lourie technique also prevents the trainer bending down. The trainer can take the stool to various situations around the farm to reinforce the lessons in different environmental situations.

Starting to work

Pups of good working strains will start to work around 6 months of age. Some strains may not start to show an interest in stock until 9–12 months old. Keeping the dog to that age is considered too expensive by some trainers. It is also accepted that very few dogs will reach a reasonable working standard if they are not working by 1 year old. Pups that mature and start to work early can be overworked before they are physically well-developed. A balance has to be achieved in allowing the pup to work to avoid boredom but not to overwork and cause physical injury. Do not set too difficult a task so that the young dog always accomplishes what is asked.

The Border Collie which is found all round the world characteristically shows 'eye' when it stares at any moving animals around the farm. This trait is also associated with a desire to gather or 'head' sheep and these two traits are expressed from 6 months of age in good strains. These dogs may be referred to as eye or heading dogs. Herding is popular with pups, and some trainers use a group of ducks for early lessons. The pup will show a desire to go around or 'head' the animals, watch them (eye them) carefully and anticipate their movements. Watching another dog work may stimulate the pup. Other strains of working dogs will start to chase stock and bark at them.

At this stage firm but kind control over the young dog is required to prevent it from chasing stock or vehicles. Temperament starts to be shown. The ideal temperament is found in a dog that is neither too hard nor too soft in its nature, and certainly not sulky. It should be calm, steady and eager to please. Close observation is needed to ensure that early defects are cured quickly. For example, strong reprimands will generally cure biting unless it is strongly inherited in the strain of dog. However, it should be realised that a dog's faults generally lie with the trainer. Then the best solution is to offer the dog to some other trainer before having it destroyed.

Once regular light training has started, the main points of the programme are:

- Establish a regular daily routine and keep to this unless the pup goes stale or sour on its work.
- Do not exhaust the pup; work it for short periods within its physical limits.
- Use a standard, repetitive procedure to help you and the dog develop mutual respect and understanding.

Further training

Although there are many different methods of training, depending on the type of dog[11], the work expected of it and the farming system, the basic needs are to teach the dog to act on command, to stop, start, go right, go left, come forward and go back. Once under command, however, the dog should show natural ability. This should not be checked.

The simplest way to teach a dog these movements is by using a 4 m long light rope or cord and a stick for restraint. The latter is a basic tool of most stock handlers; the dog should be accustomed to it and not fear it. The dog should be stroked with the stick and become used to it moving around its body. The stick is never used for hitting the dog. The dog starts to learn its routine while on short lengths of the rope, and then as proficiency increases, it is allowed more and more rope. It can then be set free trailing the cord. The check cord impedes the dog slightly, and it can be easily stepped on by the trainer to stop the dog. When the dog is under control, remove the cord. Care is needed with the rope, however, as dogs vary greatly in their acceptance of it.

Basic routines are these:

- Check that the dog is proficient at the 'sit', 'stand' or 'lie-down' command, whichever is required, and the 'get-in' or 'heel' command, so that it will walk beside you or at the heels of a horse.
- Associate the dog's name with every command for when it joins a team of other dogs. In a team of dogs, the name is a valuable individual alerting signal to warn an individual that further instructions are coming. Indeed, some well-trained, experienced dogs know what is needed without a command, and act simply on a name call. This can become a nuisance in a dog and should be checked.
- Check that it is proficient at the 'stand' or 'stay there' command. This is essential and very hard for a young, keen dog to accept, but it is vital that it remains where it is placed while other problems are attended to. Heading dogs are usually keen enough to 'stay' if there is something to eye. Some trainers warn against letting eye dogs spend too long eyeing single sheep, believing that it builds up tension. The dog should be kept moving.
- Teach the dog to come to you on command, such as 'here', 'come here' or 'come up'. A proficient dog is taught to come quickly on the command 'here' and step-by-step on the 'walk up' or 'walk on'. These routines are useful in shedding (or sep-

arating off individual sheep) and can be taught by coaxing or lightly pulling the dog towards you on the rope.

- 'Go-right' and 'go-left' responses can be taught with a small group of steady sheep, well-used to dogs. An older dog is often kept at hand to gather the small group and hold them together, if necessary. The young dog will usually cast or move around the sheep and the trainer should always stand opposite the dog, with the flock in between (fig. 53A). When the trainer moves to the right, the sheep will move to the left and the dog will respond to hold the sheep by moving to its right (fig. 53B). The dog still ends up opposite the trainer holding the sheep. When the trainer moves to the left, the sheep move to the right, and the dog moves to its left to arrive opposite the trainer again in a response to hold the sheep (fig. 53C). As the dog moves (right or left) at the response of the trainer, a word command is taught, then a whistle. Examples of word commands (which are not universally consistent) are 'come away' for GO RIGHT, and 'come bye' for GO LEFT. They may be shortened to 'away' or 'bye' or 'in' and 'behind'.

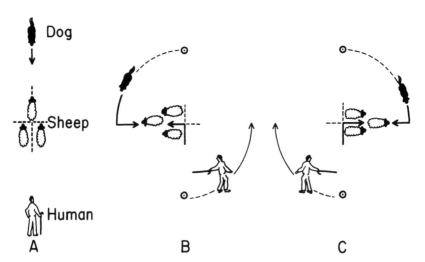

Figure 53 Teaching a dog to move right and left. A, dog holding sheep opposite handler; B, dog goes to its right; C, dog goes to its left

- Pointing the arm and the stick along with the command is useful as the dog learns hand signals for working at long distances when commands are not clearly heard.
- The dog can be taught to 'walk up' by the trainer moving backwards and allowing the dog to drive the sheep towards him. This is a good opportunity for the dog to be given practice in the

'go-right' and 'go-left' routines it has learned, as well as the 'stand' commands. Sheep that have been used a few times in this training routine soon learn that the best way to relieve the pressure of the dog is to follow the man, and this encourages the dog to follow the sheep. After a while, such sheep become a nuisance and stay too close to the trainer and are easy to fall over. They need changing frequently if this happens.

- Keep the lessons short, the commands simple and do not confuse the dog or yourself at this stage. Decide on simple commands that you can remember in crises. If the dog becomes bored or sour, stop and return him to his kennel. Always fuss over the dog after a training session. If you lose your temper, return the dog to his kennel.

- Once the dog is keen on work, take special care to teach him the 'stop-work' command. This can be very hard for a dog that is 'having fun', to learn and accept.

  The command 'that'll do' is the usual word command, and should be followed with calling the 'here' whistle as you walk away. If the dog becomes obsessed with eyeing the sheep and will not shift or leave them on command, it may have to be led away quietly by the trailing rope. Be kind and patient with the dog after you have restrained it otherwise the command 'that'll do' will come to mean trouble.

  A useful trick with an over-keen and often boisterous dog is to slow it up by putting a front paw inside its slightly loosened collar. The leg should be changed at regular intervals. Once the dog has steadied down and is under control, the leg should be released. The collar must not be too tight.

- After the dog can be commanded to stop work, it can be taken further and further away from the sheep and let go with the command to gather them. A voice command such as 'get-away' or a sound such as 'ssh-ssh', or a whistle can be taught at this stage.

- Patience is needed if a dog is not keen to leave the trainer to head or gather the sheep. Up to this point, so much of the programme has been with the dog disciplined near the trainer, that it may be frightened to leave him. The urge to 'gather in' sheep will usually start most dogs off; otherwise exciting the dog a little or causing sheep to break away may enliven their interest. With encouragement this problem is usually only short-lived.

- Do not worry about minor errors. Get the dog working and retain its interest and keenness in work.

Some trainers who want absolute control on every movement of a dog for competitive trials teach the dog its movements on a swinging arm device. Here the dog is tied to one end and the trainer rotates the other end. It is pivoted on a central post. This device can frighten the dog and is not recommended for training dogs for general work.

Refinements to training

Casting

This is a dog's ability to go around and gather sheep, and appears to be a direct behavioural carry-over from the co-operative hunting techniques of ancestral wolves. The style with which this is done varies with the dog's inherent ability and training. Some strains of Border Collie have a very wide circular cast, while the ideal aimed for in New Zealand is a pear-shaped cast. (See fig. 54.).

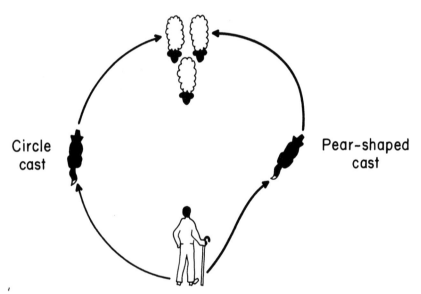

Circle cast

Pear-shaped cast

Figure 54 Direction dog takes in a circular and a pear-shaped cast

A young dog may develop a favourite side on which to cast, and it is advisable to let it work on that side until it has gained confidence in gathering, and holding sheep, and 'lifting' or starting to bring them to the handler. The dog should be started (short heads) with the sheep close at hand and then the distance should slowly be increased.

If a dog always casts on the right, it can be taught the opposite outrun by holding sheep against a fence. This forces it to go around in a left-hand direction to gather them. The sheep should be held hard up against the fence or wall so that there is no chance of the dog driving 'across the head' at the last minute and going around on its favourite side.

Teaching a dog to 'keep-out' or take a wider cast is a refinement needed by dogs that run straight at sheep, especially when the sheep are far away and out of sight. Unless the dog casts widely it will miss them. This is particularly important on broken hill country when casting blind, i.e. when sheep are out of sight. To correct this, some trainers move across the dog's run, moving the dog out wider with an outstretched stick, and one trainer[11] used a system of ropes around posts to pull the dog out wider on its outrun.

Some dog trial men teach their dogs direction by using a harness and driving it with reins like a horse. This is not recommended for general working dogs as the dog can be easily frightened. Other trainers achieve directional control by fastening a spare collar to which a cord is attached round the dog's loin. The cord is pulled to direct the dog to 'keep out' or 'keep in' using word commands. Care is needed with these techniques and experienced trainers who use them should be consulted.

Re-casting

In this routine dogs, after gathering one group of sheep, are trained to be re-directed by whistle or word command such as 'go back' to go away and collect other sheep. A good dog will go back 'on the blind', i.e. in the direction indicated even if no sheep can be seen.

This routine is taught after the dog responds well to the 'go-away' signal. This can be reinforced by an arm or stick signal pointing in the direction the dog should go. In the early stages make sure the dog can see sheep where you have re-directed it. Only when fully schooled should it be sent 'on the blind' or where there may not be any sheep.

Driving

Dogs with a strong desire to gather often find driving sheep away from the handler hard to learn. This is best taught in a well-fenced laneway or race, with a group of sheep that fill the lane, making it hard for the dog to get ahead of them. The trainer can walk along with the dog, directing it by using right- and left-hand signals, and then gradually staying back and encouraging the dog to go forward and continue the work on its own.

Occasionally, let the dog go and head the sheep and return them to you. If it has to 'get through' the fence to do this or 'over the wall', it is good extra experience, and these commands can be taught, along with arm direction.

Leading

This skill is very useful in holding back the animals in the front of a mob when droving. The dog restrains the leaders of the mob.

As the mob approaches, the dog goes further away and waits until the mob gets near. This process is then repeated and stops any animals breaking away. It is a modification of the heading skill and is taught by forcing a large mob of stock onto the dog so that it will back off at the last minute. The command 'back-off' can be taught. Much depends on the dog's nature. It should be strong enough to hold stock when directed but soft enough to back off at the last minute of its own accord.

Leading can be taught to most heading dogs but only when they are thoroughly skilled in the 'come-up' and 'back-off' routines. The 'come-behind' command can be used when you want the dog to start and work along the side of the mob to release a blockage or hurry them along.

Shedding

A dog is used to cut out, separate or 'shed' a single animal from a group. Skilled dogs will do this on their own, once the sheep to be shedded are indicated. Young dogs should be trained to do this on a 'come-here' command. The dog is placed on one side of a small group of sheep and the handler on the other. With the sheep slowly drifting between man and dog, the handler moves forward with stick outstretched to enlarge the gap between the sheep, while the dog is encouraged to 'come-up' quickly through the gap. A keen, agile dog with plenty of eye learns this very quickly. Let the dog do the work, otherwise it will never learn the finer points of the routine.

Facing up

Working dogs are expected to 'face-up' to their work. This means that they should look at the sheep directly and in confrontations with them, hold their ground or give way without 'turning tail'. This is considered to be a fault in dog trials. The aim is to teach the dog to 'back off' or move backwards, and then start a new approach.

Dogs can be taught to face up while on a check-cord using a stick to direct them or with a collar around the loins controlled by a cord. They can also be taught to walk backwards on command. One trainer held his dog between two short boards and physically pushed it back on command. Whatever method is used, great care is needed to reassure the dog and avoid fear.

Teaching when to bite

Biting can be a major fault in heading dogs. Generally it has to be stopped in the very early stages of training by severe reprimand. However, sometimes a dog is required to use its teeth when attacked by a ewe or cow. To teach this, a close-up confrontation between the dog and some old ewes or wethers (not rams) in a small yard should be arranged. Encourage the dog to walk up on the sheep and then 'stand'. When a sheep attacks excite the dog (e.g. by giving a hissing sound) so that it bites to defend itself when charged. Repeat the exercise until the dog gains confidence.

There is no fear that this will make the dog into a biter, provided that the 'stop' control has been properly taught in its early training.

Catching sheep

Dogs that will catch ewes and lambs are very useful, especially at lambing time. They generally grab the ewe by the wool near the head and hang on until the shepherd can catch it. They may tip the ewe over and then hang on. With lambs, good dogs 'mouth' the lamb across the neck or hold it down with their front feet. Shepherds differ widely in their views as to how much encouragement or training a dog should be given to catch sheep because of the risk of encouraging worrying. From the wide range of opinion, the following points emerge:

- Dogs should not be encouraged to catch sheep until they are under complete command.
- Only dogs with suitable temperaments should be encouraged, and this is best judged after their basic schooling. Dogs with a tendency to bite should not be trained to catch sheep.
- Pups should not be encouraged to catch by playing with them using a piece of string on a stick trailing some sheepskin or a lump of rubber. This can start bad habits.
- A dog can be encouraged to catch sheep by exciting it and grabbing hold of a single sheep by the rear end. The dog will generally come to your assistance and grab the sheep's front end. The

command 'hold it' or 'catch it' will soon be learned. Once the dog has a correct grip on the sheep, it can be let go for the dog to hold on its own. Vigilance is needed to call the dog off if any injury is likely to happen to sheep or dog.

Heeling

Most strains of Border Collie with plenty of eye will also 'heel' cattle. Huntaways also do this but this trait is best seen in the Australian Cattle Dog or Heeler bred especially for the purpose.

Unless the dog learns the correct technique of biting the beast low down near the hocks of the back leg, while it is on the ground, and then lying low before the beast kicks, it will be severely injured. Some dogs bite indiscriminately and swing on tails which is not the technique required. Good cattle dogs must also 'nose' cattle by running along biting the beast's nose while it charges at them. The aim is to get the dog to turn the animal then heel it to hurry it along.

It is essential that the dog is fearless but under strict command so that its actions are controlled by the stockman. There is a risk that a team of cattle dogs may get out of control, some heeling and some nosing causing stress and chaos.

Huntaways

'Huntaway' dogs are used (in New Zealand) for general stock handling and mustering. They are barking dogs. Some strains will 'head' or gather sheep as well as 'hunt' or drive sheep away from them; some bark all the time they are moving, while other strains can be more easily trained to bark on command.

Huntaways can be readily trained to bark by clapping hands or shouting to excite them in the presence of stock. The command 'speak up' is used, with a whistle command. The stop or 'shut up' command is taught by raising the voice at the pup or growling and shaking it when you want silence.

Most Huntaway breeders and trainers insist that the first requirement of a dog is to be able to gather or head sheep. If it cannot do this it is very restricted in its general purpose duties.

Huntaways are then taught to drive or hunt sheep both towards or away from the trainer.

The teaching routine is similar to that outlined for the heading dog, remembering that the Huntaway does not have the same inherent cast or stalk as the Collie. The training routine is to teach 'stand', 'speak up', 'walk up' and so on, as the dog goes away from the handler. The trainer can accompany the dog at first during the routine, and then hang back as the dog goes forward towards the stock. Eventually the dog with keenness will go on its own for long distances.

Jumping

The dog should be taught when very young to jump by lifting and encouraging it to jump on to or over any obstacle it meets, such as the kennel roof, wool bales, trailers and the decks of trucks, motorbike seats and so on. It should be commanded to 'get up' or 'get up there' with a hand signal pointing to the obstacle. Dogs soon learn to balance on moving objects like motorbikes if lifted on and reassured for short steady journeys before full speeds are attained. One breeder taught his dogs to balance by giving them short trips in a wheelbarrow covered with a sheepskin.

Backing

Huntaways are taught to jump on sheep's backs by lifting a young dog on to the backs of woolly sheep (not newly shorn). The sheep should be tightly packed in a pen or race. The dog should be petted and reassured, and then left to walk off the sheep over the rails and out of the pen. It can then be encouraged to jump into the pen onto their backs to the command 'get up there'. Only when the dog is competent to walk on the sheep's backs should it be put into pens of loose sheep as should it fall off, it may get stood on and attacked by some sheep at close quarters. This can frighten some dogs. Reassurance by the handler is important at this stage.

Negotiating fences

Unless a dog can get through a fence, its usefulness is very restricted. Problems arise because there are so many kinds of wire fence, and some of them are either partly or completely electrified.

Pups should be exposed to fences early. Rearing pups in cages often prevents them learning how to get through fences.

With non-electric fences, dogs that do not learn themselves can be quickly taught by pushing them through a large space, using the command 'go through'. Later, an arm direction can be added. Dog biscuits can attract a young dog through but plenty of reassurance usually is all that is needed. Once the dog learns that it will not get caught, it will squeeze through smaller and smaller spaces between wires and through netting.

Some trainers insist that dogs should be taught to crawl below the bottom wire to avoid being caught. Others insist that dogs should be taught to jump clearly over the top. This is acceptable for young, active dogs but as they age and get heavy, they tend to put their feet on the top wire and can get torn on barbed wire, or become entangled.

Electric fences

Electric fences pose special problems mainly because the 'hot' wires are not always at the same height on the fence. Many dogs which are shocked will not go near the fence again, no matter how much reassurance is given to them. Some suggestions are:

- Try to be consistent in placing 'hot' wires on a fence, so that either the bottom or middle wires are not electrified. The dog can be taught where the safe area is.
- Insist that the dog 'comes through' the fence to you, hoping that it will learn to jump after the initial shocks. This may not work.
- Teach the dog to crawl through the fence at a low point to avoid 'hot' wires higher up the fence. This could be done by placing its feed and water below a clearly marked electrified wire. The dog would have to learn to crawl to obtain its feed.
- Low electric fences (30–50 cm high) can be jumped with no difficulty, but the dog may get a shock through the ground as it approaches the fence, so it has to learn to take a very long leap.
- It may be possible to have safe, non-electrified access points that could be recognised by the dog, but which sheep would not learn to use.

Whatever method is used, depending on the type of fence and the dog, it is vital that the dog never associates the shock of the fence with the trainer. Some trainers insist that when the dog is shocked they do not even speak to the dog in case this happens. The dog can be reassured by the trainer a short interval after the shock.

Buying a trained dog

Problems may occur when a trained dog changes ownership. The task is first to establish a good relationship with the new dog and this is best done by taking it all round its new environment while on a lead, generally reassuring it at regular intervals, and especially at feeding times. Let it off in a confined area to make sure it will stay with you. Do not start working it until you are sure it will do so.

Most dogs should work after about 7–10 days with the new owner if they are keen dogs.

Problems can arise with word commands (perhaps in a different language or dialect) and with new-sounding whistles. If possible, the new owner should obtain a tape recording of the previous owner's commands, and try to imitate them until the dog gets used to them in modified form from the new owner. It may take some time for the dog to respond fully to these new sounds.

One breeder trains the dogs to respond to a referee's whistle with the pea removed, and uses Morse code.

The commands are:

—   One long, to call attention.
– –   To bark or 'speak up'.
— – — Steady, and when repeated, sit down.
— —   Go away. To change direction, one long —, then point or move across, then two long — — again.
— —   Come in behind.

The advantage of this technique is that anyone can work the dog, and this is often important in emergencies. With other whistles, there is generally too much room for personal variation in the notes blown, and the dog only recognises his master's note. The loyalty of one-man dogs can perhaps be marvelled at but there are great practical drawbacks.

Great care and patience are needed with a new dog, recognising the anxiety it is feeling as a result of the change of environment, especially if it has previously been a one-man dog.

Working a team of dogs

Stockmen vary greatly in their methods with a team of dogs. The main concern is to prevent the team working as an uncontrolled pack. The prime aim is to insist that dogs only work when commanded. The simplest and best way to achieve this is to use the dog's name as an 'alert' signal and then have a different stop whistle for each dog. After that all the commands for each dog are the same.

Good control of a team is achieved by regular work and firm but kind control. Using different commands and whistles for each dog can become very confusing, especially for anyone else who has to work the dogs in an emergency.

Discipline

All trainers agree that discipline is essential in training and that any reprimand must be given as soon as the crime is committed. Even a few seconds delay will be ineffective. If possible the dog should be caught and stopped in the act. The punishment should also fit the crime.

Many working dogs are beaten out of all proportion to the crime but this is often an outlet for the frustration and stress of a tired, overworked stockman. A loud shout, a growl, a shake are generally all that is needed, and in a team of dogs, shepherds can teach the dominant dog in the group on the command 'shake him' to go and attack the rule-breaker. Often the lower ranking dogs will assist. Care is needed not to abuse this technique. This is useful, as the criminal receives the treatment immediately.

Electronic collars which the handler can use to give the dog a shock from a distance via radio control are decried by many experienced trainers and are banned in some countries. Some trainers use them only for serious faults such as night barking, biting or sheep worrying. The main point is to teach the dog that the shock is another form of the trainer's severe disapproval before it is administered. This is done close at hand by growling, shaking or hitting the dog, along with the shock. Then at a distance, the dog associates the shock with disapproval. Otherwise it will not know where the shock came from or what it was for.

It is generally only used on hard dogs that have developed extremely bad habits which cannot be cured by other means. All trainers stress that if faults are recognised and checked instantly, there is no need for electric collars.

Instant discipline at a distance from the handler for example for a strong-willed dog that always forgets its schooling, or that breaks the rules only when out of the handler's reach, can be ad-minstered by the sharp crack of a stock whip or a starting pistol. Also, some trainers are very skilled at throwing small stones which must be on target for best effect. However dogs soon learn to ignore these tricks so they have to be changed. The main point to remember after reprimanding a dog is not to release it immediately, but to hold it in a submissive position until it lies quietly. After a while the dog can be reassured to re-establish a normal relationship before it is released.

Other tasks for dogs

There appears to be no limit to the tasks various breeds of working dogs can be trained to do. Some examples are:

- Guide dogs for the blind[12,13].
- Finding people buried in snow, landslides and earthquakes.
- Locating buried mines and bombs[1].
- Sniffing out concealed drugs.
- Tracking criminals and lost people; crowd control and apprehending criminals in the Police Service.
- Guard dogs in security services.
- Guard dogs to live with sheep and protect them against predators such as coyotes[10].
- Gun dogs to mark or point the position of game, and dogs to retrieve shot game.
- Dogs to detect cows in oestrus in indoor barns. Accuracy has been shown[9] to range from 73−90% for individual dogs.
- Obedience competitions and scenting competitions.
- General hunting of game and vermin.

Problems with working dogs
Behavioural problems in dogs are well-described[3] in terms of their cause and methods of correction. The following comments refer to those of special concern in the working dog.

Biting people
Experience from veterinary practice[3] has shown that most people-biting problems occur when highly excitable and aggressive pups are placed in environments that over-stimulate their active defence reflexes. The main situations where biting occurs are:

- The newly whelped bitch, in defending her litter from strangers.
- To contest the top dominant position with the dog's owner or trainer.
- To protect a run or kennel from a stranger.
- In response to removal of food.
- In fear of being handled by law-enforcement officer or veterinarian.
- In response to being teased by children or excessively beaten by their owner.
- Through confusion and frustration, when a dog receives different, inconsistent orders from different people. One person may praise and another reprimand for the same act, such as coming into the house or the front of the truck.

Apart from the first, where strangers should keep away from bitches with pups, the solution to these problems lies in correct socialisation and handling of the pup in the critical 6−10 week stage. In particular, emphasis should be given to handling the dog's mouth. Care should always be taken when approaching strange dogs. It is wise to stand still and let the dog come to you. Hold out a hand and let it sniff you, and then touch it on its chest rather than on the top of its head or shoulder. Grabbing the dog's shoulder is a very dominating and threatening act and may elicit aggression in dogs that do not show the normal submission postures when you approach.

Veterinarians advise that dogs which start biting people can rarely be stopped with any guarantee that the problem will not arise again under stress. Therefore, such dogs are not to be trusted, and should be disposed of.

Biting sheep

This problem can only be dealt with in the early training stages where the dog clearly learns that it is not allowed to bite except on command. It is considered a major fault in a working dog, although it must be realised that strong-eyed heading dogs have an inherent ability to hold their quarry after the stalk and will bite. Mature dogs that consistently bite and injure stock can be muzzled during work but care is needed to ensure this does not restrict breathing when working in hot conditions. For valuable dogs, some veterinarians will remove the canine teeth. However, other veterinarians refuse to remove teeth or cut them off because of the risk of injuring the dog's jaw and causing permanent pain in the sawn-off teeth. Sheep can still be bitten by a dog with canines removed. Dogs which are inherent biters should be disposed of.

Sheep worrying

Working dogs of any type are all capable of becoming sheep killers if they are allowed to develop this habit. Whether they are fed on meat or not has little to do with the problem. They often develop the habit through being allowed to roam in combination with another farm dog, such as an eye dog and a terrier. Poor vision may also start dogs biting sheep and from there they start worrying.

Working dogs should always be tied up when not working, and owners should always know where their dogs are. The sudden disappearance of a dog should always be investigated. In some countries (e.g. New Zealand) the law requires working dogs to be tied up when not working for animal health reasons.

Bolting

This is where a dog leaves its work and runs away, usually home to its kennel to await its very annoyed master. Beating the dog at that stage is no solution, as the problem is caused by insufficient elementary training. It is often the result of excitement and fear of the owner shouting at it or at another dog.

One trainer uses a long nylon line on the dog's collar, tied around a short stick. When the dog bolts, the line invariably becomes entangled with some object. The farmer leaves the dog there while continuing to work near until its interest to return to work is enlivened. The dog is not severely reprimanded. New dogs may bolt before they have formed a friendly relationship with their new owner. Keep a new dog on a check cord until you are sure it will stay with you, especially when you start shouting.

The best approach to this problem is after the dog bolts the first time, check what the problem was and whether the dog's behaviour was justified. If the dog bolts more than once, then view it as a serious problem which will be hard to cure. Getting rid of the dog may be the best solution.

Car sickness

Dogs may vomit, defecate and urinate when travelling in vehicles. This can be avoided by getting the dog to sit in the vehicle with a lot of reassurance and petting, then going very short distances, stopping before the dog is sick, and again providing reassurance. The distances travelled can then be slowly increased and lavish praise usually solves the problem. Efficient anti-vomiting drugs are now available.

Leg mounting

Working dogs often develop the habit of mounting and thrusting on people's legs. Knocking them off sometimes makes them more keen to continue. Some male dogs also learn to mount sheep, especially while being taught to jump on their backs.

The solution is actively to discourage such behaviour in the playful puppy stage. Divert the pup's attention and admonish it when it tries to mount. However castration may be the only sure solution.

Sniffing people

Dogs that sniff the crotch of visitors can be embarrassing for the owner and very worrying for the visitor. It is often a habit of big, friendly Huntaways, and when the stranger pushes them away from the front, they quickly go around and sniff the rear, causing the person great embarrassment. The cure is actively to discourage pups from starting this habit, especially with any women or girls in the household. Divert the dog's attention to other play or work. A good reprimand will stop a young dog doing this or jumping up with dirty feet at strangers. Standing on the dog's back toes when it jumps up is often an effective cure.

Scent rubbing

Dogs often roll in strong offensive-smelling materials such as rotten carcasses and food, faeces from other dogs and farm animals, as well as afterbirths. They rub their heads, cheeks, throats, shoulders and backs in the material and usually scratch the ground near the object with their hind feet when they are finished.

It has been suggested[5] that rolling in the material may reduce the novelty of it by the smell mingling with the animal's own smell. It may also cause more social investigation of the animal by others in the pack, which may be a comforting experience to the dog.

Some individual dogs seem to be more keen on scent rubbing than others, and the only cure seems to be to check it in the young pup when it starts the habit by a reprimand and diverting its attention.

Feral dogs

Feral dogs are becoming an increasing problem as more and more dogs are neglected, abandoned or escape from urban homes. Few behavioural studies of the problem have suggested practical solutions. It appears from studies[14] that feral dogs showed pack behaviour similar to wolves but did not vocalise or carry food to their pups. This latter may limit their population growth. They did not show aggressive behaviour to humans until they were trapped.

The danger is that feral dogs as an intelligent species can quickly learn to modify their behaviour and become problems in farming areas.

Welfare

From the information available on dogs, the following are some recommendations to help owners provide good care while getting the best from their working dogs:

- Punishment must be given immediately the crime is committed. It should be administered in its mildest form and there is no place for cruelty and brutality.
- Positive acts of petting and body touching are adequate rewards for most dogs.
- Clean, dry, draught-proof housing and runs are essential for working dogs. They should not be left to roam while not working.

- Dogs feed infrequently and diets must meet their nutritional needs. Overfeeding leads to obesity and disease.
- Dogs not working need exercise.
- Pups should be adequately socialised to other dogs, people and their environment from 10 weeks of age.
- Dogs should be taught by reassurance how to handle the many hazards found on farms.
- Physical mutilations such as castration, tail docking and de-barking are not needed in working dogs. Removal of dew claws is often very necessary.
- Breed standards may cause unacceptable physical and behavioural changes which would prevent such dogs being used on farms[18].
- Training is essential in farm dogs but it should not become an end in itself. Excessive training may be the result of an owner's misdirected ego, and critics who see this consider that the dog has had some of its rights removed.

This chapter has only dealt with the farm dog and information on other dogs and their problems is given in the list of further reading.

References

1. Ashton, E.H. and Eayrs, J.T. (1970) 'Detection of hidden objects by dogs.' In *Taste and Smell in Vertebrates*, eds G.E.W. Wolstenholme and J. Knight, Sympos. Ciba Fdtn J.A. Churchill: London, pp. 251—63.
2. Bleicher, N. (1962) 'Behaviour of the bitch during parturition.' *J. Amer. vet. Med. Assoc.* 140(10), 1076—82.
3. Campbell, W.E. (1975) 'Behaviour problems in dogs.' *Illinois: Amer. vet. Publ. Inc.* 306pp.
4. Fox, M.W. (1966) 'Dog development, rearing and training. Practical applications of research findings.' *S.W. Vet.*, Summer issue, 303—5.
5. Fox, M.W. (1968) 'Socialisation, handling and isolation.' In *Abnormal Behaviour in Animals*, ed. M.W. Fox, Philadelphia: W.B. Saunders Co., p. 563.
6. Hart, B.J. (1974) 'Normal behaviour and behavioural problems associated with sexual function; urination and defecation.' *Vet. clinics Nth. Amer.* 4(3), 589—606.
7. Hopkins, S.G., Schubert, T.A. and Hart, B.L. (1976) 'Castration of adult male dogs: effects on roaming, aggression, urine marking and mounting.' *J. Amer. vet. Med. Assoc.* 168, 1108—10.
8. Igel, G.J. and Calvin, A.D. (1960) 'The development of affectional responses in infant dogs.' *J. Comp. Physiol. Psychol.* 53(3), 302—5.
9. Kiddy, C.A., Mitchell, D.S., Bolt, D.J. and Hawk, H.W. (1977) 'Use of trained dogs to detect oestrous-related odours from cows.' *Proc. 69th Meet. Amer. Soc. Anim. Sci. 1976* (abstr), 176—7.

10. Linhart, S.B., Sterner, R.T., Carrigan, T.C. and Henne, D.R. (1979) 'Komondor guard dogs reduce sheep losses to coyotes. A preliminary evaluation.' *J. Range Mgmt.* **32**(3), 238–41.

11. Mills, A.R., Herbert, S.F. and McIntyre, M.V. (1964) *A Practical Guide to Handling Dogs and Stock*, A.H. & A.W. Reed, Wellington, New Zealand, 125pp.

12. Peel, B.W. (1980) *Proc. Sem. on Dogs, Univ. Waikato, N.Z.*, ed. L. Sissons, Univ. Waikato Cont. Educ. Dept., pp. 28–33.

13. Pfaffenberger, C.J. and Scott, J.P. (1959) 'The relationship between delayed socialisation and trainability in guide dogs.' *J. Genet. Psychol.* **95**, 145–55.

14. Scott, M.D. (1971) *The Ecology and Ethology of Feral Dogs in East-central Alabama.* Ph.D. Thesis, Auburn University, Alabama, 213pp.

15. Scott, J.P. (1971) 'Barking in dogs.' In *Acoustic Behaviour in Animals*, ed R.G. Busnel. New York: Elsevier, pp. 25–8.

16. Seager, S.W.J. and Platz, C.C. (1978) 'Semen collection in the inexperienced male dog.' *Appl. Anim. Ethol.* **4**(1), 94 (Abstr.).

17. Vollmer, P.J. (1978) 'Puppy rearing: 2 Establishing the leader–follower bond.' *Vet. Med. Small Anim. Clin.* Aug. pp. 994–7; Sept 1135–6.

18. Woodhouse, B. (1970) *Dog Training my way*, London: Faber and Faber.

Further reading

Burns, M. and Fraser, M.N. (1966) *Genetics of the Dog*. London: Oliver and Boyd.

Carlson, D.G. and Griffin, J.M. (1982) *Dog Owners Own Veterinary Handbook*. Howell Book House, pp. 1–364.

Fox, M.W. (1965) *Canine Behaviour*. Springfield: C.C. Thomas, pp. 1–137.

Fox, M.W. (1972) *Understanding Your Dog*. New York: Coward, McCann and Geoghegan, 240pp.

Kelly, R.B. (1958) *Sheep Dogs*. Sydney: Angus and Robertson, pp. 1–237.

Longton, T. and Hart. E. (1969) *Your Sheep Dog and its Training*. Battle (Sussex); Alan Exley, pp. 1–76.

Lorenz, K. (1953) *Man Meets Dog*. London: Penguin Books, pp. 1–198.

Scott, J.P. and Fuller, J.L. (1965) *Genetics and Social Behaviour of the Dog*. Chicago: University of Chicago Press, pp. 1–468.

Tortora, D.F. (1977) *Help! This Animal is Driving me Crazy*. Chicago: Playboy Press, 306pp.

Wolfensohn, S. (1981) 'The things we do to dogs.' *New Scient.* **90**(1253), 404–7.

10 Handling and Welfare

Improved management, handling and welfare of livestock using current behaviour knowledge

This final chapter attempts to draw together some of the points mentioned in the chapters on each species and some general points to help stockmen use current behaviour knowledge to improve handling and overall welfare of the stock. This is to meet the requirements of the 'domestic contract'.

Handling and handling devices

Handling has been defined[16] as 'the process in which man or his substitutes carry out an operation on a domestic animal, which demands some restraint or compliance by the animal involved'.

Examples of a substitute for man may be a dog, leader sheep, ridden horse, farm motorbike, helicopter, electric prodder, crowding gate, a conveyor belt and so on. Handling devices used for animals are yards, pens, stalls, fences, walls, crushes, chutes, halters, harness and so on which are used to control the animals' reactions or escape. The amount of restraint given to animals during handling may vary from a little (e.g. moving stock to a new field) to complete restraint (e.g. in foot trimming). Animals may also be forced to crowd together which infringes their individual space needs.

Stress

Stress has been defined in a variety of ways. Selye[36] calls it 'the non-specific response on the body to any demand made upon it'. A useful mnemonic has been developed[1] of Stress = Situations That Release Emergency Signals for Survival. 'Stressors' are events outside the animal which stress it and the final effect on the animal is 'strain'.

There are many ways proposed to measure stress in animals such as thyroid hormones[7]; corticoids[11,17,27,32,38]; catecholamine levels[28,41,42]; blood glucose[23,33]; changes in heart or respiration rate[30,31] ; or somatic cells in milk[44]. Behavioural signs of stress include 'displacement activities' where for example animals in the middle of a fight may stop to graze[13,34] or stereotypic responses such as the circular walking of animals around their cage[24]. Distress is a term used to describe the damaging or unpleasant aspects of stress, some of the latter being needed each day to keep animals alive and attuned to their environment and its dangers. Boredom is an unpleasant aspect of stress and occurs where normal stress levels have diminished to a point where there is insufficient stimulation of the animal which then becomes lethargic and almost comatose. Boredom can be as harmful as an overload of different stressors.

The effect of many stressors

Animals can usually tolerate the effect of a single stressor over a short period without undue affect on production. However, if the animal has to suffer a number of stressors at the same time, health and performance are affected and death may result. Trouble often arises because some of the disease stressors may be sub-clinical and not seen (e.g. mastitis or internal parasites).

A stockman is often faced with a dilemma as to whether, when he has the flock or herd in the yard, they should be held to complete a whole range of jobs (e.g. injections, drenching, tagging), or whether they should be brought in separately for each job. In the latter case there may be more stress through the extra mustering. Another example is where day-old chicks are put into large houses with fluctuating temperatures, after several injections, beak trimming and no knowledge of where feed and water can be found. Doing all these management routines at the same time has great practical advantages for the farmer but may over-stress the birds.

There is currently little information to guide stockmen with the answer to these problems.

Stress at critical times – birth

One of the most critical times in an animal's life is at the birth of its offspring. As young animals form a large proportion of the profit from an enterprise, then inevitably the stockman wishes to take some part in the supervision of the event. This may save many offspring but it often leads to overzealous interference and stress to the animal. As the capital value of the animals increases, so does the stockman's concern and desire to assist.

Where the dam is milked, the offspring may be weaned soon after birth and the dam returned to a milking group and forced into a new routine. This can cause great distress.

The move to 'easy-care' or minimal shepherding of lambing ewes in New Zealand is an example of a husbandry change by farmers who believe that they cause more lambing trouble by close shepherding than they do by leaving their ewes undisturbed. There is clearly a happy medium between the extremes of too much interference and attention at birth, and neglect on the other. The stockman must decide this level based on the behavioural and economic aspects of the enterprise.

Animal reactions to handling

The reactions of animals to specific handling situations vary greatly and stockmen talk of animal 'temperament'. Tables 8 and 9 give a number of temperament scores for cows and bulls by different authors[5,9,39] as examples. The chief weakness of these scores is that milking cows can react quite differently to the different people who milk them. A cow rated 'skittish' by one milker can be 'placid' for someone else[14]. Nevertheless, milking temperament is an important trait that has been improved by selection.

Table 8 Suggested categories for temperament ratings of cows (after Tulloh[39]; Dickson[5])

Tulloh's ratings	Dickson's ratings
1. *Docile.* Cow stands quietly, may raise or lower head; may lean forwards or backwards. Well positioned for observations; in general is unperturbed.	1. Very quiet, never gives any trouble; extremely docile during milking and preparation; the ideal milker.
2. *Slightly Restless.* Generally docile but moves frequently; i.e. every few seconds. Flicks tail and blows quietly through nostrils. Stubborn docile animals are included here.	2. Stands quietly in stall; not bothered by preparation for milking but may move frequently shifting weight from side to side; may flick tail occasionally; gives very little trouble.
3. *Restless.* Moves about continuously pulling or pushing on the side of the crush. Stance difficult to define, frequently flicks tail, snorts, difficult to see ear tag; may be stubborn.	3. Generally quiet, but moves around a lot; may lift feet occasionally during preparation or milking but does not kick. Flicks tail frequently or appears restless at times.

Table 8 (*contd*)

Tulloh's ratings	Dickson's ratings
4. *Nervous.* Restless animal which quivers when hand is on its back. Defecates in bale – may also be stubborn.	4. Appears very restless during preparation for milking; kicks at handler occasionally. Steps from side to side a great deal. Quivers when a hand is placed upon her.
5. *Wild.* Restless; struggles violently; bellows and froths at the mouth – may be stubborn.	
6. *Aggressive.* Wild animal; attacks observer by kicking. Attempts at biting or lunging – may also be stubborn.	

Table 9 Suggested categories for temperament ratings of bulls (Fraser[9])

1. Stable personality showing little self-will and much interest and curiosity. These bulls, usually sexually immature, show compliance when being handled (49% of sample).

2. Stable personality, mature sexually. Show some sex drive and are masterful with some self-restraint. They are audacious without being malevolent (22% of sample).

3. Of uncertain stability; sometimes aggressive and generally of high dominance; i.e. show aggressiveness to other bulls; high in libido (8% of sample).

4. Very aggressive even to man, who is threatened or attacked. However, frequency of such acts is not high (5% of sample).

5. High in libido and aggressiveness to man (potential killer bulls). Exhibit constant aggression (5% of sample)

6. Apprehensive bulls, many of them exceptionally fearful. Always taking flight from disturbing stimuli (11% of sample).

Studies of dog temperament have shown that some (e.g. terriers) are more aggressive than others such as beagles that make ideal laboratory animals as they live happily together in groups. Strain differences in temperament in poultry are also recognised and this has important implications in multiple-caged birds where space is restricted and food may not be on offer at all times.

Reducing distress in handling

The Brambell report[3] considers that in intensive units, injuries to animals are the greatest source of pain. It is also declared[2] that the major task of welfare is to 'minimise stress, discomfort and unease of farm animals before they become acute'. This is the 'art' of stockmanship.

Behaviour studies have shown that there are times and ways in which certain manipulations can be done to animals to allow them to cope better with domestic handling, and some of these are discussed below.

Using the early critical periods

Research has shown that stockmen can exploit the early critical periods in animals when attachment to people is easier and this results in easier future handling. An example is the few minutes spent stroking (or gentling) newly hatched chicks over several days (called TGC or Tender Gentle Care) which will result in quieter hens later that will even come and feed from the human hand. Untouched chicks will panic and they risk being smothered. TGC has also been shown to make animals more able to resist infections and even starvation.

Another example is handling foals soon after birth[43] which does not upset the mare–foal bond but makes later handling much easier. Hand-reared lambs are a well-known example of how early bonds with humans last until maturity and even after these animals are returned to the flock.

In later-developing species like the dog, the critical socialisation period is from 4 to 6 weeks after the eyes and ears are open. Before day 10, no such socialisation is possible. The benefits of early socialisation to assist training and avoid later problems are well recognised. The converse of this is also well recognised by the severe damage done to animals badly handled when young.

Using the old to teach the young

Offspring learn a great deal about their environment from their parents. This is especially the case in 'follower-type' species that learn the tracks, food supplies, watering points and so on from parents. How much of this learning–teaching is lost by early weaning of offspring is not well documented. Early weaning is becoming an increasing practice in farming to increase output from the dam, the best example being in the pig. The problem of getting day-old chicks to eat and drink exists because they cannot learn from the tidbitting of their dam.

There are many more examples in the previous chapters, probably the best being work in Australia[12] which shows that the best way to teach lambs to eat drought feed is to give it to them when they are still with their dams. With newly farmed species like deer, allowing the young to learn from well-managed older animals is vital for ease of handling.

On the reverse side, it must be remembered that older animals can also quickly teach bad habits to the young. Examples of this are crib-biting in horses and sheep worrying in dogs.

Conditioning before stressful events

The opportunity should always be taken where possible to condition animals to an event before it happens. A good example is long-distance travel where giving the stock a number of short trips in the days before the journey is very useful. Studies[19] have shown that on a 5-day journey, calves learned to travel better as each day progressed, and were much quicker to load and unload at each new stopping point.

The importance of conditioning or pre-training is also seen in getting new heifers used to the milking facilities, and stock can also be conditioned to drops in winter temperatures so that energy costs of heating houses may be reduced.

Allowing animals time to make their own pace into unfamiliar territory rather than forcing them is always worthwhile. It is not always possible however and certain species like pigs with poor vision often give problems. Drugs and sedatives are used in some countries to stop pigs fighting during transit. Fights occur mainly when vehicles are stopped.

Some events, like slaughter, are only experienced once by animals, and it is currently believed that they cannot be anticipated in well-run modern slaughtering facilities, hence stress is minimal and suffering not present. Distress can also be caused by frustrating 'conditioned' animals as seen in delayed milking. Some cases of animal neuroses occur as a result of unexpected events or from being forced to make a choice where no clear alternatives are available. 'Learned helplessness' can be set up where established routines are changed and animals are punished regularly.

Capitalising on the animal's ability to learn

Most farm animals have a considerable capacity to learn new things and yet farmers who are prepared to spend time training working dogs and horses do not exploit the learning ability of other species. The use of trained leader sheep[4] is an example of a worthwhile project for sheep farmers where labour is limited and flock size increases.

However, attitudes are changing and there are now some good new examples of animals' learning ability being used to improve husbandry. These are where animals drink from nipples, use special dunging areas, press buttons to regulate heat[25] as seen in pigs which are quick to learn and generalise widely. Hens also learn quickly but do not generalise well. Hens can learn to control ventilators and activate emergency feed and water systems by pecking. When given a choice, hens prefer to work for their food[22]. Channelling part of an animal's effort into useful work reduces problems that arise from vices or other anomalous activities. More research is needed in this area.

Recognising animals' sensory capacities

Each farm species uses its senses in a particular way and it is important to realise that these are quite different to humans. The challenge to the stockman is to appreciate what each species can see, smell, hear and feel in order to anticipate their reactions, especially during handling. Some general points are:

- *Sight* Farm animals have good panoramic vision but more limited binocular vision. Blindfolding animals is an established technique to pacify them during transit if they become distressed. In the head-down grazing position, animals with eyes on the sides of their head see all round, but when the head is raised there is a clear blind spot behind the head and down the muzzle.
- *Hearing* Farm animals generally have good hearing. Hens hear sounds up to 8000 Hz while cattle can hear sounds over 18 000 Hz which is above the range of human hearing. It is well known that dogs can hear high-pitched whistles that humans cannot hear. There is little information on low frequency sound detection but it is known that farm animals often hear sounds which are not perceptible to the farmer.

  A major concern is the noise levels in modern intensive poultry and pig houses. These can reach over 80 decibels which can harm human hearing. Little is known of their effects on the animals.
- *Vocal calls* Each species has its own particular calls for different

situations but few stockmen are skilled enough to understand or imitate them to use them during husbandry. Some managers claim to be able to use the sounds heard outside intensive houses to determine the state of the animals inside, but this has not been adequately researched. Animal alarm calls can certainly be used by experienced stockmen to avoid problems and their meaning should be learned by trainees.

- *Smell* A good sense of smell is common to all farm animals but is particularly well developed in dogs and pigs. Smell is the main way offspring are identified with the exception of the hen where it is done by sound. The importance of pheromones in oestrus and mating is now well recognised and the sensitivity of animals to their own dung has important husbandry implications. This is seen where animals will not graze around their own dung patches but it also means that dung can be used as a repellant to prevent bark-chewing of trees.

- *Taste* Animals eat their traditional preferred feeds with few problems but taste preferences become important when wastes or other by-products are offered to them. Preferred feeds may not lead to greater intake but disliked feeds can reduce intake dramatically. The assessment of whether some of the proprietary feed additives do really influence the preferences of animals needs to be fully researched.

- *Touch* Stockmen have traditionally used touch or 'gentling' to quieten animals during handling and this is especially recognised in the horse and dog. However, it should be recognised and used more in the other species. Animals regularly invite others to groom them and they spend time grooming or preening the parts of their bodies that they can reach. Manipulating and stroking animals can cause some of them to move into a type of hypnotic trance. This can be done to hens and some dogs. Some horse trainers are said to use hypnosis as an aid in early handling.

- *Body posture language* An animal's moods are often displayed by body postures and these can be used by stockmen to improve understanding and ease handling. Body posture covers such things as the position of the head, condition of the eyes and placement of ears and tail. It is also concerned with the fight or flight stance (in bulls) and dominant or submissive postures. Little of this information is currently documented so stockmen must gain it by experience.

The way species meet is important for stockmen to appreciate. Some meet nose to nose (naso–naso) like the horse and pig, while others like the dog meet nose to tail (naso—ano). The greeting technique of blowing up the nostrils of horses is often used by trainers to reassure their animals.

Remember that animals have good memories. It has been shown that cattle and sheep have retained information about handling procedures for at least 3 years. They especially remember experiences of bad treatment.

- *Group size* As man has changed the natural social grouping of farm animals, so too have some aspects of their development and amenability changed. This is seen most readily in sheep where in general, four sheep in a group are required for the expression of typical sheep-like behaviour. Isolating a sheep from the flock can cause distress[17], and it should be recognised that five animals/treatment group are a minimum for valid research results if these results are to be useful for flock sheep.

It has also been demonstrated[21] that if new-born kids are separated from their dams at or near birth, their subsequent behaviour is restricted and they often die before maturity.

Putting large numbers of animals of the same sex together is a common farming practice but it may lead to abnormal behaviour problems. Examples are the incidence of 'buller steers' in large groups of steers in feedlots, or of severe pecking in hens when roosters are absent.

The stockman should aim to keep stable groups together for as long as possible and try not to introduce strangers to them. Even temporary removal of an animal may alter its social order when it is returned to the group. Young animals introduced to mature groups inevitably go down to the bottom of the social order and this greatly affects their performance.

The relentless chasing of single animals that have broken away from a group should be avoided. It causes great stress and often injury as they break fences in panic. Other animals should be released to join them, or they should be ignored and left to re-join the group of their own accord. Chasing single animals on farm motorbikes can cause great stress.

Stockman experience in handling

During the restraint and handling of stock, people often get hurt. Arms and legs are the most vulnerable parts for injury from kicking or biting animals. Horses can cause most injuries in New Zealand[18] followed by cattle, but dogs and sheep also take their toll.

In approaching stock, experienced handlers use the point of balance of the animal to move them. In side view this is behind the shoulder and from the front it is the centre of the head as shown in fig. 55. Moving to increase pressure (by going closer) or to relax pressure (by moving back) on one or other side of this point will make the animal move in the required direction. The point of balance can also be demonstrated by sitting astride a sheep. Apply pressure to the head and it will reverse while pressure to the rear will make it go forward.

Behind shoulder

Head centre

Figure 55 The point of balance behind the shoulder and in the centre of the head of an animal (after Kilgour and Houston (1979) — see ref. 18, page 279)

The use of short lengths of plastic pipe to extend the arms is useful for drafting animals as is a flag on the end of a 1 m stick. Electric prodders should be avoided where possible as they can stress the animal and cause confusion. Beating animals may be good therapy for an angry farmer, but it does little to accomplish what is required of the animal. There are good texts available[6,8,20] giving details of correct animal restraint.

Designing facilities to reduce stress

The design of handling facilities is vital to successful animal handling where the work must be done with minimal stress or injury to both humans and beasts. However, facilities are often inadequate because they have been designed from the humans' view and the animal's much lower viewpoint, especially when it lies down, has been ignored. There are certain areas where animals benefit by seeing others, and other situations where seeing other stock may be a distraction. Distractions up to 500 m away from the parlour can affect cows during milking.

It is good that debate continues about the shape of handling facilities and design of farm buildings as this leads to controlled investigations. The fact that some species prefer to eat at the same time (e.g. hen, pig, cattle) must be appreciated when new buildings are designed, as well as the fact that space around an animal's head may be more important than space measured in ground area. Good examples are the feeding devices for cattle into which they put their heads and battery cages for hens.

Pen shape is important as seen where cattle preferred a triangular-shaped pen to a square design as they could settle and not encroach into each other's space while feeding[37].

Loading and unloading ramps for animals are a special area of concern as they can greatly increase distress, especially on the way to slaughter. The result can be severe bruising of carcasses and death. Engineers are now working with ethologists to improve designs and this must be encouraged.

It is a good exercise for farmers[35] or consultants to check on animal handling and housing facilities by noting down in a balance sheet both the positive and negative factors at work. A milking parlour is a good example where such things as sharp turns, electrified crowding gates, fluctuating vacuums, old cup liners, the number of sticks around the yard to beat the cows with and so on, can be noted.

Positive recommendations can nearly always result, such as adjusting rail heights, roughening floors and so on. It is wise to visit other units or invite other farmers to visit yours and make suggestions. Little of this information is available in published form.

Transport

Farm animals are usually transported at some time in their lives and may travel long distances by road, air and sea. This can cause large changes in their environment through continuous motion, long periods of waiting, variations in climate and being re-grouped with other unfamiliar animals. As a result, concern is often expressed about welfare during transit and in many countries special legislation has been enacted to protect animals at this time.

The following are some points to consider in animal transport so that the stress may be reduced and the basic welfare needs met.

- Avoid mixing strange animals for transport. Aim to transport them in groups that have previously been together.
- Before transporting, animals should be carefully conditioned to the feed they will receive on the journey. They can also be conditioned

to the actual pens and given short trips to allow them to gain experience.

- Animals from outdoor pastures generally cope well with temperature changes and transport stress.
- Animals raised indoors however often find the cold, wind and movement stressful and they can suffer great transport shock. Attempts should be made to protect them from these stresses.
- Ships carrying livestock from temperate zones through the tropics require appropriate ventilation systems that will meet the animals' needs, especially for stock housed below deck.
- For animals transported by air, especially horses, straps should be fitted to prevent them losing floor contact during turbulence.
- Cattle should be de-horned before transit to prevent damage to their mates and their handlers.
- Calves travel better in a truck in large pens than in small pens. They try to avoid the sides of the pen.
- Adequate head room, especially on lower decks is essential to allow stock to stand normally and not get down and be trampled.
- Tranquillisers can be used under veterinary supervision. These are often used for very valuable animals, or those travelling without the company of mates. The sedative can be tried some days before travel to check the dose rate. There is wide variation in the reaction of individuals to drugs.
- Meat animals being transported to slaughter need special care to avoid stress and bruising which will downgrade the carcass.
- The floors of transport vehicles must be suited to the class of stock to allow them to get adequate grip without damaging their feet. Insecurity causes great distress.
- Space must be adequate for the comfort of the animal. If animals are too loosely packed, they will be thrown about and if too tight any animal that gets down will be trampled, or smothered.
- Animals should be checked at regular intervals during the journey. Rest periods should be provided and special care is needed to ensure adequate ventilation and temperature control for stock held in vehicles when they are stopped.
- Although animals may not fight during travel itself, fighting can erupt when vehicles are stationary.
- On long journeys by road or rail adequate feed, water and rest periods must be given.
- After transport, animals must be given adequate food and water and be allowed to rest. They will also spend time investigating their new environment.

A good summary of behaviour and the stresses of transport is given in references 10, 26, 29 and 40. When legislation is formulated to cover the needs of animals during transport, it is important to make sure that the literal application of the law does not go against the welfare and comfort needs of the animal that it was designed to protect.

Pre-slaughter handling

Some farm animals are grown for slaughter, while others need to be slaughtered near the end of their productive life. Although many people are quite happy to exploit a domestic animal as a pet or for their own recreation and sport, they feel uneasy when it comes time for its death. Very strong feelings about so-called 'humane' slaughter techniques fill the literature and many of the agencies for the protection of animals have arisen out of concerns over animals near and at slaughter. It is not sensible or logical to pair the words 'humane' with the word 'slaughter'. Methods of final slaughter are not more or less humane. It is true even when 'humane slaughter' is reserved for the actual manner of dispatch i.e. electrical stunning, captive bolt pistols, gassing with carbon monoxide or carbon dioxide, Kosher slaughter and so on.

The fact that each new hygiene and transport regulation requires animals to be held for long periods at abattoirs which constitute alarming environment for already stressed animals to cope with, does not appear to concern those propounding the doctrine of humane slaughter. In some countries sheep after arrival at abattoirs have to be washed, dried overnight and finally reach the slaughter point some 24–35 hours after arrival. This is justified by the need for hygiene, despite the fact that meat is a cooked product and that no statistics can be cited for human deaths from eating meat directly contaminated from the conditions of slaughter. Not only is hygiene not an issue, but the effects of stressors and even strain, plus the physiological insult of a death by gassing or electrical stunning rather than simple knife-cut slaughter, results in tough meat from stressed animals.

The pale soft exudative meat from pigs and the dry dark cutting meat from cattle are both direct results of the quite unwarranted long holding period, mixing of strange animals and general consequences of inappropriate regulations which currently surround the slaughter of farm animals. The task of rendering an animal insensitive or unconscious before its blood is let (which is by definition being humane) is not that simple. Sheep skulls are harder than cattle skulls.

Rams back off for head-to-head fighting clashes while cattle bunt more gently. Mechanical impact stunning of sheep is a different problem to impact stunning cattle. In some cases, especially with calves, it is now known that the body is paralysed by the electric shock while the animal remains fully conscious of pain, and is unable to struggle at all. This was the case with some of the electric boxes used to put down animals like dogs in Britain for years. The animals were paralysed and suffered greatly until as a consequence of the stopped heart beat, they finally died some 26 seconds later. This was all done in the name of being humane.

Death to animals should be carried out preferably in the known environment, among flock or herd mates in the simplest and quickest manner. The emotive request by some people that sheep should be killed out of sight of their mates completely ignores their species-specific needs for companions, even at slaughter.

It is sad to have to conclude that the stresses which surround the slaughter of farm livestock are very serious and are almost without exception the result of agitation from people of misguided goodwill, or bureaucrats who are imposing regulations related to some mythical requirements for better meat hygiene.

To be humane to animals[15] is (i) to handle them in a manner least distressing to them and (ii) with full consideration for their normal species-specific behaviour requirements.

Final comment

The final chapter of this book has sketched some of the practical guidelines to help stockmen to handle animals with humane concern and skill. The art of stock handling has not progressed far towards being a science, but some of the areas of better restraint and handling procedures are being studied. Readers of this book can themselves make a contribution to this area and the authors would be pleased to hear of any experience gained with any species. Millions of domestic animals in towns and country need not be in discomfort if these skills and findings are properly communicated to others.

References

1. Amoroso, E.C. (1967) *Environmental Control of Poultry Production*, ed. T.C. Carter, Edinburgh: Oliver and Boyd.
2. Brambell, F.W.R. (1965) *Report of the Technical Committee to enquire into the Welfare of Animals under Intensive Livestock Husbandry Systems.* London, HMSO, Cmnd, Paper No. 2836, 85pp.

3. Brambell, F.W.R. (1967) Address to Society for Veterinary Ethology, *Proc. Soc. Vet. Ethol.* **1**(3), 1−4.

4. Bremner, K.J., Braggins, J.B. and Kilgour, R. (1980) 'Training sheep as leaders in abattoirs and farm sheep yards.' *Proc. N.Z. Soc. Anim. Prod.* **40**, 111−16.

5. Dickson, D.P. (1967) 'A study of social dominance, temperament and learning ability in dairy cattle.' *Diss. Abstr.* **28**, 402B.

6. Ewbank, R. (1968) 'The behaviour of animals in restraint.' Chpt 10 in *Abnormal Behaviour in Animals*, ed. M.W. Fox, London: W.B. Saunders, pp. 159−78.

7. Falconer, I.R. and Hetzel, B.S. (1964) 'Effect of emotional stress and TSH on thyroid vein hormone level in sheep with exteriorized thyroids.' *Endocrinol.* **75**, 42−8.

8. Fowler, M.E. (1978) *Restraining and Handling of Wild and Domestic Animals*, Iowa: Iowa State Univ. Press, pp. 1−332.

9. Fraser, A.F. (1957) 'The disposition of the bull.' *Brit. J. Anim. Behav.* **5**, 110−15.

10. Hails, M.R. (1978) 'Transport stress in animals: a review.' *Anim. Regul. Stud.* **1**, 289−343.

11. Holcombe, R.B. (1957) 'Investigation on the urinary excretion of "reducing corticoids" in cattle and sheep.' *Acta Endocrinol. Suppl.* **34**, 1−96.

12. Keogh, R.G. and Lynch, J.J. (1982) 'Early feeding experience and subsequent acceptance of feed by sheep.' *Proc. N.Z. Soc. Anim. Prod.* **42**, 73−5.

13. Kilgour, R. (1969) 'Animal behaviour under stress.' *N.Z. J. Agric.* **118**(1), 44−7.

14. Kilgour, R. (1975) 'The open-field test as an assessment of the temperament of dairy cows.' *Anim. Behav.* **23**, 615−24.

15. Kilgour, R. (1978) 'The application of animal behaviour and the humane care of farm animals.' *J. Anim. Sci.* **46**, 1478−86.

16. Kilgour, R. (1978) 'Minimising stress on animals during handling.' *Proc. 1st Wld Congr. Ethol. appl. Zootechn. Madrid*, E-IV-5, pp. 303−22.

17. Kilgour, R. and de Langen, H. (1970) 'Stress in sheep resulting from management practices.' *Proc. N.Z. Soc. Anim. Prod.* **30**, 65−76.

18. Kilgour, R. and Houston, J.B. (1979) 'Handling animals safely.' *N.Z. Frm.* **100**(18), 157−8.

19. Kilgour, R. and Mullord, M. (1973) 'Transport of calves by road.' *N.Z. Vet. J.* **21**, 7−10.

20. Leahy, J.R. and Barrows, P. (1953) *Restraint of Animals*, 2nd edn, Ithaca, N.Y.: Cornell Campus Store Inc.

21. Liddell, H.S. (1954) 'Conditioning and emotions.' *Scient. Am.* **190**, 48−57.

22. Matthews, L.R. and Kilgour, R. (1978) 'Free-operant studies on farm animals and their relevance to intensive farming.' *Proc. 1st Wld Congr. Ethol. appl. Zootechn. Madrid*, E-I-6, pp. 71−6.

23. McKay, D.G., Young, B.A. and Thompson, J.R. (1972) 'Glucose metabolism in sheep: acute cold stress.' *J. Anim. Sci.* **34**, 900 (abstr.).

24. Meyer-Holzapfel, M. (1968) 'Abnormal behaviour in zoo animals.' In *Abnormal Behaviour in Animals*, ed. M.W. Fox, Philadelphia: W.B. Saunders, pp. 476−503.

25. Morris, G.L. and Curtis, S.E. (1981) 'Operant supplemental heat for swine nurseries.' *J. Anim. Sci.* **53**, suppl. 1, 188 (abstr.).

26. Moss, R. (ed.) (1982) *Transport of Animals intended for Breeding, Production and Slaughter*, The Hague: Martinus Nijhoff Publishers, pp. 1–236.

27. Panaretto, B.A. and Ferguson, K.A. (1969) 'Pituitary adrenal interactions in shorn sheep exposed to cold, wet conditions.' *Aust. J. Agric. Res.* **20**, 99–113.

28. Pearson, A.J., Kilgour, R., de Langen, H. and Payne, E. (1977) 'Hormonal responses of lambs to trucking, handling and electric stunning.' *Proc. N.Z. Soc. Anim. Prod.* **37**, 243–8.

29. Pearson, A.J. and Kilgour, R. (1980) 'The transport of stock – an assessment of its effects.' In *Behaviour in Relation to Reproduction, Management and Welfare of Farm Animals*, eds M. Wodzicka-Tomaszewska, T.N. Edey, and J.J. Lynch, Armidale: University of New England Press, pp. 161–5.

30. Pearson, R.A. and Mellor, D.J. (1975) 'Some physiological changes in pregnant sheep and goats before, during and after surgical insertion of uterine catheters.' *Res. Vet. Sci.* **19**, 102–4.

31. Pearson, R.A. and Mellor, D.J. (1976) 'Some behavioural and physiological changes in pregnant goats and sheep during adaptation to laboratory conditions.' *Res. Vet. Sci.* **20**, 215–17.

32. Purchas, R.W. (1973) 'The response of circulating cortisol levels in sheep to various stresses and to reserpine administration.' *Aust. J. Biol. Sci.* **26**, 477–89.

33. Reid, R.L. and Mills, S.C. (1962) 'Studies on the carbohydrate metabolism of sheep: XIV. The adrenal response to psychological stress.' *Aust. J. Agric. Res.* **13**, 282–95.

34. Schein, M.W. and Fohrman, M.H. (1955) 'Social dominance relationships in a herd of dairy cattle.' *Brit. J. Anim. Behav.* **3**, 45–55.

35. Seabrook, M.F. (1973) 'A study of the influence of the cowman's personality and job satisfaction on the milk yield of dairy cows.' *J. Agric. Labour Sci.* **1**, 49–93.

36. Selye, H. (1974) 'The implications of the stress concept.' *Biochem. Expt Biol.* **11**, 190–9.

37. Stricklin, W.R., Graves, H.B. and Wilson, L.L. (1979) 'Some theoretical and observed relationships of fixed and portable spacing behaviour of animals.' *Appl. Anim. Ethol.* **5**, 201–14.

38. Thurley, D.C. and McNatty, K.P. (1973) 'Factors affecting peripheral cortisol levels in unrestricted ewes.' *Acta Endocrinol.* **74**, 331–7.

39. Tulloh, N.M. (1961) 'Behaviour of cattle in yards. II. A study of temperament.' *Anim. Behav.* **9**, 25–30.

40. U.F.A.W. (1975) 'Transport of farm animals.' *UFAW Sympos.* held 1974, pp. 1–32.

41. Van der Wal, P.G. (1972) 'Influence of electrical stunning on pigs for slaughter – changes in blood.' *Dtsch. tierarztl. Wschr.* **79**(5), 97–120.

42. Von Euler, U.S. (1964) 'Quantification of stress by catecholamine analysis.' *Clin. Pharmcol. Therapeutics.* **5**, 398–404.

43. Waring, G.H. (1970) 'Primary socialization of the foal (*Equus caballus*).' *Amer. Zool.* **10**(3), 25 (abstr.).

44. Whittlestone, W.G., Kilgour, R., de Langen, H. and Duirs, G. (1970) 'Behavioural stress and cell count of bovine milk.' *J. Milk Fd. Technol.* **33**, 217–20.

Appendix 1: Definitions of Welfare

A selection of definitions from the literature

- 'A wide term that embraces both the physical and mental well-being of the animal[2].'
- 'Existence in reasonable harmony with the environment, both from an ethological and physiological point of view[4].'
- 'Welfare can be satisfied if three questions can be answered in the affirmative: (a) Are the animals producing normally? (b) Are they healthy and free from injury? (c) Is the animal's behaviour normal[1,9]?' Similar definitions to above[6].
- 'Welfare is a relative concept. Profit is a matter related to welfare and determining the *relationship* between welfare and profit is a scientific matter. The *choice* between welfare and profit is an ethical matter[3].'
- Considering that while an animal is producing protein without observable signs of pain, then it can be considered to be comfortable. 'Distress, strain, gross abuse and suffering are used to describe unfavourable circumstances ... the presence of raw flesh or heavy bruises would be classified this way, but callouses or hard skin caused by concrete floors and producing no pain would not be classed as distressing[8].'
- 'On a general level it is a state of complete mental and physical health where the animal is in harmony with its environment. On an empirical level it may be measured by studying an animal in an environment which is assumed to be ideal, and then comparing it with an animal in the environment under investigation[5].'
- 'Handling animals in the least disturbing manner with full consideration for their normal species-specific behaviour requirements[7].'

References

1. Adler, H.C. (1978) 'Animal welfare in farm animal production.' In *The Ethology and Ethics of Farm Animal Production*, ed. D.W. Folsch, Basel and Stuttgart: Birkhauser, pp. 13–17.

2. Brambell, F.W.R. (1965) (Chairman) 'Report of the Technical Committee to Enquire into the Welfare of Animals kept under Intensive Livestock Husbandry Systems', London: HMSO, Command Paper No. 2836, pp. 1–85.
3. Brantas, G.C. (1975) Quoted in 'Introduction to plenary session I. Intensive confinement farming: Ethological, veterinary and ethical issues.' *Proc. 1st Wld Congr. Ethol. appl. Zootech, Madrid. 1978*, pp. 25–6.
4. Report of Committee of Experts on Animal Husbandry and Welfare (1977): Policy recommendations, The Hague: Dutch National Council for Agricultural Research, pp. 1–37.
5. Hughes, B.O. (1976) 'Behaviour as an index of welfare.' *Proc. 5th Europ. Conf., Malta*, pp. 1005–12.
6. Jackson, W.T. (1976) 'Design and farm animal welfare.' *Vet. Rec.* **99**, 64–6.
7. Kilgour, R. (1978) 'The application of animal behaviour and the humane care of farm animals.' *J. Anim. Sci.* **46**, 1478–86.
8. Randall, J.M. (1976) 'Environmental and welfare aspects of intensive livestock housing: a report on an American tour.' *Nat. Inst. Agric. Engin. Silsoe, Rept.* No. 19, pp. 1–54.
9. Simonsen, H.B. (1977) cited by Adler, H.C. (1978) *The Ethology and Ethics of Farm Animal Production*, ed. D.W. Folsch, Basel and Stuttgart: Birkhauser, pp. 15–17.

Appendix 2: Definitions of Intensive and Factory Farming

Intensive farming has been variously defined as follows:

- 'The system in which the farmer uses technology with discretion and although livestock numbers are increased, the pattern of life the animals lead does not significantly change[2] .'
- 'More animals per unit, less space per animal and mechanical equipment instead of persons attending animals[1] .'
- A delicate compromise between conflicting priorities: capital costs, production costs, output, efficiency of labour and welfare[3] .'

Factory farming has been defined as:

- 'An extreme restriction of freedom, enforced uniformity of experience, submission of life processes to automatic controlling devices and inflexible time-scheduling . . . and running through all this is the rigid and violent suppression of the natural[2] .'

References

1. Adler, H.C. (1978) 'Animal welfare in farm animal production.' In *The Ethology and Ethics of Farm Animal Production*, ed. D.W. Folsch, Basel and Stuttgart: Birkhauser, pp. 13–17.
2. Harrison, R. (1977) *Animals in Factory Farms*. Information Report: Animal Welfare Institute, Washington, **26**, 1–3.
3. Hughes, B.O. (1973) 'Animal welfare and the intensive housing of domestic fowl.' *Vet. Rec.* **93**, 658–62.

Appendix 3: Viewpoints on Welfare

The welfarists

People who are mildly or strongly opposed to many or all forms of intensive husbandry or factory farming. Their criticisms are usually focussed on the following points:

- Confinement — the animal is deprived of space, freedom, sunshine and fresh air.
- Automation — no human contact, no human examination of its condition, no animal companionship.
- Rough handling — in transport, loading, slaughter, plus excess noise.
- Mutilation — in castration, beak trimming, tail removal, restricted lighting regimes.
- Disease — greater risks as stocking density increases. More drugs needed and development of resistance to drugs, with repercussions for humans.
- Pollution — of the human environment by wastes, excreta and smells.

The farmers or operators

People who are interested in or gain a livelihood from farming animals. The main points put forward by farmers can be summarised as:

- Farmers would not be farmers if they disliked (and were cruel to) animals.
- Animals that produce well must be content and free from stress. (This has not been accepted by the German Welfare Act or the Brambell report — see reference 2.)
- Farmers working with animals are likely to understand them best and know when they are suffering.

- Farmers must work within the economic constraints put upon them by markets and consumers. The changes proposed by welfarists would increase prices and consumers generally refuse to pay more for them.
- Farmers cannot afford to respond to short-lived emotive fads but would accept changes recommended by independent and knowledgeable bodies.
- Animals in unprofitable systems are at a very great welfare risk.

It is important to realise that farmers differ considerably in their attitudes to various intensive systems and their desirability. The National Farmers Union in Britain has presented farmers' views in a pamphlet[6].

The public

The public generally is not interested in welfare unless attention is focussed on certain issues by the news media. The main issue seems to centre around the question as to whether urban consumers will pay more for products produced by 'welfare-approved' systems. Generally they will not. An example was where 80% of Swiss people voted by referendum in favour of banning certain poultry cages. In one region at least, only 20% would pay the 7% extra price required to cover the change of system.

There is always the added concern that the chosen standards of the vocal affluent minority may force changes which the less affluent majority cannot afford. The public should retain the right to information, to inspect and have free expression of opinion on all aspects of animal farming.

The researcher

Researchers have a clear commitment first to obtain objective information on animal behaviour, and then to publish it and make it available as widely as possible.

One very experienced scientist presented a good summary of research workers' views[4]:

- There will always be cruel handling by some operators and coddling by others, regardless of the law or fixed or official standards.
- To define cruelty in terms of dimensions of living quarters is to adopt a mindless approach to both husbandry and to the policing of husbandry methods.

- It is the skill of the man and the welfare of the stock which are important, not merely a certificate of competence.
- The amount of space is not as critical as the design; a space may be cruel, comfortable or boring; the difference lies in its design.
- Codes of practice are better and more flexible than legal regulations in setting and upholding husbandry standards.

The animal

A range of carefully controlled experimental approaches is now available to ask animals questions. For example some workers[1,5] allowed trained pigs to decide on air temperatures and others[3] allowed cows to show preferences between feeds by pressing buttons to raise feed buckets. Techniques are also available for animals to decide on floor types, to measure the effect of change in pen design and levels of electric current felt by cows[7].

Thus if research into behaviour is not neglected, those who have the welfare of farm animals at heart will be able in future to design environments and equipment, based on sound facts provided directly by the animals about their needs and preferences.

References

1. Baldwin, B.A. and Ingram, D.L. (1967) 'Behavioural thermoregulation in pigs.' *Physiol. Behav.* 2, 15−21.
2. Ewbank, R. (1969) 'Behaviour implications of intensive animal husbandry.' *Outlook on Agric.* 6, 41−6.
3. Matthews, L.R. and Temple, W. (1979) 'Concurrent schedule assessment of food preference in cows.' *J. Exp. Anal. Behav.* 32, 245−54.
4. McBride, G. (1976) 'Improving the welfare of livestock.' *Proc. 53rd Ann, Conf. Aust. vet. Assoc.*, pp. 7−9.
5. Morris, G.L. and Curtis, S.E. (1982) 'Let the pigs turn on the heat.' *Pig Internat.*, Jan. issue, pp. 30−32.
6. N.F.U. (1980) National Farmers Union Public. U.K.
7. Whittlestone, W.G., Mullord, M., Kilgour, R. and Cate, L.R. (1975) 'Electric shocks during machine milking.' *N.Z. Vet. J.* 23, 105−8.

Appendix 4: Approach to a Practical Behaviour Problem

The following is an actual example of a practical problem that applies to countries with a national artificial insemination programme with cattle. While bulls are waiting for progeny-test data before extensive use, they are usually run together in groups. It is too expensive to keep them tethered singly for this 2–3 year 'lay-off' period. There is a high injury rate among these pastured bulls. Why? The problem is analysed in a series of steps.

(1) Identify the problem
Examination of the veterinary records and stockman's notes will show that the injuries to bulls are either short-term and minor, or long-term and permanent. Injuries are more likely at a particular stage and certain bulls seem to be injury-prone.

(2) Literature search
A search of the literature in a science library will provide any relevant information already known about the problem. This can provide advice on techniques and methods used by previous investigations. It also warns of problems found by others.

(3) Inspection of the unit
The unit should be inspected to find out the regular routines and to talk to the staff. Attention should be given to the paddock sizes, stocking rates, group sizes, breeds, age-distribution of animals within each grazing group, movement routines, yard designs, stockmen's methods and any other details considered important.

Before a behaviour study is planned, discuss it with the staff to gain their co-operation.

(4) Decide what data to record
From inspection of the unit and staff interviews, the likely behaviour problem areas will be predictable. For example:

- Investigative: approaching other bulls, grooming, licking, lip-curl (flehmen).
- Agonistic: threat, pawing, bellowing, throwing up dirt, digging holes, sparring in pairs, fighting in groups, head-to-head pushing.
- Sexual: chasing, rear-mounting, head-mounting, attempted mounting.
- General: dispersal patterns over the paddock, time and place of social interactions and severity of encounters should be recorded.

These are the relevant sections of the cattle ethogram that should be used.

(5) Design the study

Advice should be obtained from experienced behaviourists where possible and statisticians experienced in the analysis of biological data so that valid conclusions can be drawn from the data. Some possible suggestions could be:

- Observe three groups: young bulls (2–3 years); middle aged (3½–4½ years) and old bulls (over 5 years).
- Choose paddocks with good vantage points and divide arbitrarily into grids. The paddocks could be pegged out, photographed from a vantage point to act as reference points for the grid and the pegs removed.
- Observations made during daylight hours by two separate observers. Two full-day observations taken for each group of bulls, each 3 days apart.
- Be able to identify the bulls by markings — natural or artificial so that they can be recognised at a distance.
- Ask the staff to keep the management as routine as possible during observations.
- For special management functions such as yarding, consider this as a separate study, taking the same analytical approach as outlined above.

(6) Analysing the results

Actual results are shown in table 10. The results would be made up of:

- The social behaviour of each bull.
- Herd behaviour score every 30 minutes.

- Photographs taken every 30 minutes to show the position of each bull by superimposing the paddock grid on the photograph.
- Each observer's independent records examined for agreement and hence reliability.
- Records of any unexpected happenings.
- Distances travelled and mob positions in the paddock.

Table 10 The number of social interactions recorded on one day in bulls of different age groups adjusted for group sizes[1]

	Sexual	Agonistic		Amicable	
Age (yrs)	Mounting	Fighting	Pawing	Grooming	Lip-curl
Young 2–3½	15	32	19	27	55
Medium 3½–4½	275	197	3	6	4
Old 5½–6½	6	53	53	2	21

(7) Drawing the conclusions

Care is needed to base conclusions on the observed facts and not on enthusiastic conjecture. If the data were unreliable, no conclusions should be drawn. From valid conclusions, practical recommendations should be made to the staff. They should also be thanked for their co-operation. From an actual study in New Zealand[1] recommendations were that steps should be taken to clear the paddocks of tree stumps and other potential hazards, water troughs should be placed near fence lines and not in the middle of paddocks, and that the middle-aged bulls at pasture should be treated with extra care. The very young bulls had a low agonistic score, and the very old were very territorial hence caused few problems except when yarded.

Reference

1. Kilgour, R. and Campin, D.N. (1973) 'The behaviour of entire bulls of different ages at pasture.' *Proc. N.Z. Soc. Anim. Prod.* **33**, 125–38.

Appendix 5: Behaviour Inventories or Ethograms

Ethograms

An ethogram is a precise and detailed catalogue of the responses which make up an animal's behaviour. The details can be most easily recorded on movie film or videotape and from this each element of the response pattern can be analysed to make up the ethogram. If these techniques are not available careful observation can be used to record the detail.

Using certain classifications[2] an ethogram for sheep and goats was developed[1]. This is shown in table 11. These same divisions[2] can be used by readers as an exercise for any chosen farm species. Another example of an ethogram, for a range of dogs is given in reference 3.

Table 11 Behaviour patterns of sheep and goats (after reference 1)

Behaviours	Characteristic pattern
Ingestive	Grazing, browsing, ruminating, licking salt, drinking.
	Suckling — nudging udder with nose, suckling, wiggling tail.
Shelter-seeking	Moving under trees, into barns.
	Huddling together to keep off flies.
	Crowding together in extreme cold weather.
	Pawing ground and lying down.
Investigatory	Raising head, directing eyes, ears and nose towards disturbance.
	Nosing an object or another sheep.
Allelomimetic	Walking, running, grazing and bedding down together.
(groups)	Following one another.
	Bouncing stiff-legged past an obstacle together (sheep).
Agonistic	Pawing, butting, shoving with shoulders.
(conflict,	Running together and butting (rearing and butting, in goats).
fighting,	Bunching and running.
escape)	Lying immobile.
	'Freezing' (in young kids only).
	Snort and stamp foot ('sneeze', in goats).

Behaviours	Characteristic pattern

Eliminative	Urination posture:
	    squatting of females;
	    arching back, bending legs (male lambs).
	Defecation:
	    wiggling tails.

Care-giving (epimeletic) (Females only)	Licking and nibbling placental membranes and young. Arching back to permit nursing, nosing lamb at base of tail. Circling newborn lamb. Baaing when separated from lamb.
Care-soliciting (et-epimeletic)	Baaing (distress vocalisation by lamb when separated, hungry, hurt or caught). Baaing of adults when separated from flock.

Sexual (male)	Courtship: following female, pawing female,
	    hoarse baa or grumble;
	    nosing genital region of female;
	    sniffing female urine, extending neck with upcurled lip;
	    running tongue in and out;
	    rubbing against side of female; biting wool of female;
	    herding or pushing female away from other males.
	Copulation: wiggling tail (rare); mounting, thrusting move-
	    ments of hind quarters.

Sexual (female)	Courtship: rubbing neck and body against male; mounting male (rare). Copulation − standing still to receive male.

Play	Sexual: mounting (by either sex).
	Agonistic: playful butting.
	Allelomimetic: running together;
	    'gambolling' (bouncing stiff-legged and turning in air
	        together).
	'Game playing' − jumping off and on rock or log together.

Species-specific behaviour

Behaviour that is species-specific is that which is carried out in the same fixed way and not altered by learning. Other behavioural responses are flexible and change more easily with learning. When a duck takes her newly hatched brood of mixed ducklings and chickens to water, the chicks are reluctant to enter the water. Swimming is not part of the species-specific behaviour. Ducklings follow her into the water without hesitation. The ethogram catalogues the whole behaviour repertoire.

Domestication has not greatly altered the species-specific behaviour patterns of farm animals. Studies of wild or feral animals can contribute much to our understanding of farm animal behaviour and this information is sometimes the only assistance for the ethogram at present available.

References

1. Hafez, E.S.E., Cairns, R.B., Hulet, C.V. and Scott, J.P. (1969) 'The behaviour of sheep and goats.' Chapter 10 in *The Behaviour of Domestic Animals*, 2nd edn, ed. E.S.E. Hafez, London: Bailliere, Tindall and Cassell, pp. 296–348.
2. Scott, J.P. (1956) 'The analysis of social organization in animals.' *Ecol.* **37**, 213–21.
3. Scott, J.P. and Fuller, J.L. (1965) *Genetics and Social Behaviour of the Dog*, Chicago: Chicago Univ. Press, pp. 63–5.

Appendix 6: Further Reading

Alexander, G. and Williams, O.B. (1973) *Pastoral Industries of Australia*. Sydney Univ. Press, 567pp.

Allen, R.D. and Westbrook, W.H. (1979) *The Handbook of Animal Welfare*. Garland STPM Press.

Arnold, G.W. and Dudzinski, M.L. (1978) *Ethology of Free-ranging Domestic Animals*. Amsterdam: Elsevier Scientific Publishing Co.

Brambell, F.W.R. (1965) *Report of the Technical Committee to Enquire into the Welfare of Animals kept under Intensive Husbandry Systems*. London: HMSO, Command Paper No. 2836, pp. 1–85.

Broom, D.M. (1981) *Biology of Behaviour*. Cambridge: Cambridge University Press, 320 pp.

Clarke, J.A. (1981) *Environmental Aspects of Housing for Animal Production*. London: Butterworths, 511 pp.

Craig, J.V. (1981) *Domestic Animal Behaviour: Causes and Implications for Animal Care and Management*. New Jersey: Prentice-Hall Inc, 364 pp.

Curtis, S.E. (1981) *Environmental Management in Animal Agriculture*. Illinois: Animal Environment Services, 385 pp.

Folsch, D.W. (ed.) (1978) *The Ethology and Ethics of Farm Animal Production*. Stuttgart: Birkhauser Verlag, 144pp.

Fowler, M.E. (1978) *Restraint and Handling of Wild and Domestic Animals*. Iowa: Iowa State University Press, 332 pp.

Fraser, A.F. (1980) *Farm Animal Behaviour*, 2nd edn, London: Bailliere Tindall, 291 pp.

Hafez, E.S.E. (Ed.) (1975) *The Behaviour of Domestic Animals*, 3rd edn, Baltimore: The Williams and Wilkins Co., 532 pp.

Houpt, K.A. and Wolski, T.R. (1982) *Domestic Animal Behavior for Veterinarians and Animal Scientists*, Ames: Iowa State Univ. Press, 356 pp.

Kiley-Worthington, M. (1977) *Behavioural Problems in Farm Animals*. Oriel Press.

Manning, A. (1979) *An Introduction to Animal Behaviour.* Edinburgh: Edward Arnold Ltd, 329 pp.

Neufert, E. (1980) *Architect's Data — Handbook of Building Types.* 2nd edn, London: Granada.

Owen, J.B. (1974) *The Detection and Control of Breeding Activity in Farm Animals.* Aberdeen: School of Agriculture, 109 pp.

Pond, W.G. and Houpt, K.A. (1978) 'Behaviour'. Chapter 2 in *The Biology of the Pig*, Ithaca: Comstock Publishing Associates, pp. 65–81.

Sainsbury, D.W.B. and Sainsbury, P. (1979) *Livestock Health and Housing*, 2nd edn, London: Bailliere Tindall, 388 pp.

Scott, J.P. (1958) *Animal Behaviour.* Chicago: University of Chicago Press, 281 pp.

Squires, V.R. (1981) *Livestock Management in the Arid Zone.* London: Inkata Press, 271 pp.

Syme, G.J. and Syme, L.A. (1979) *Social Structure in Farm Animals.* Amsterdam: Elsevier Scientific Publishing Co., 200 pp.

Thorpe, W.H. (1966) *Learning and Instinct in Animals.* London: Methuen & Co., 558 pp.

Tomaszewska-Wodzicki, M., Edey, T.N. and Lynch, J.J. (eds) (1980) *Behaviour in Relation to Reproduction, Management and Welfare*, Armidale, NSW: University of New England Press, 202 pp.

UFAW (1979) *The Humane Treatment of Food Animals in Transit.* UFAW, Potters Bar, 43 pp.

UFAW (1979) *The Welfare of Food Animals.* UFAW, Potters Bar, 68 pp.

West, G.P. (1977) *Encyclopedia of Animal Care.* Baltimore: Williams and Wilkins, 867 pp.

Wood-Gush, D.G.M. (1971) *The Behaviour of the Domestic Fowl.* London: Heinemann Educational Books, 147 pp.

Glossary

Abattoir: premises at which animals are slaughtered.

Accessory claw: the extra non-functional claw, higher up the leg of an animal. Often called the dew claw.

Acoustic: concerned with hearing.

Ad libitum: a system of feeding where the animal has free choice and can eat until its appetite is satisfied (satiated).

Afterbirth: the membranes which carry the young animal inside the dam's uterus. It is expelled separately after the offspring is born.

Agonistic: used as a word for general aggression in animals. Covers the responses of approach, fight and flight.

Ahemeral: time intervals that do not add up to 24 hours.

Allanto chorion: allantoic fluid – the fluid around the foetus.

Allelomimetic: a type of social behaviour involving imitation of another species member or members. Behaving in the same way in a group.

Alpha male: the top-dominant male in a social-ranking order.

Altricial: young hatched or born in a very immature condition generally blind, naked and unable to leave the nest or den.

Androgenised female: a female treated with male hormones (androgens). Acts like a male in mating response.

Anogenital: the area around the anus and vagina in females or the anus and testicles in males.

Anthropomorphic: assigning to animals (or gods) the feelings and emotions of humans.

Anti-ovulatory: a drug to stop ovulation.

Artificial insemination (AI): the technique of collecting the male sperm and inserting it using a pipette into the female reproductive tract.

Bail: a frame for holding an animal (e.g. milking bail or head bail).

Barrow: a castrated male pig.

Binocular vision: the animal's ability to see with both eyes while looking ahead of it.

Birth site: the area on which an animal prepares for and gives birth to her offspring and mothers them.

Blanket insemination: insemination of all the cows in a herd regardless of their oestrous state.

Bloat: a condition where the gas from fermentation in the rumen cannot escape and the animal blows up.

Bobby calf: calf slaughtered when over 4 days old and which has been fed solely on whole milk.

Bolt: to run away, out of control.

Bolus: the round ball of grass produced in the cow's mouth after chewing and passed down to the rumen.

Bonding: see *Imprinting.*

Branding: giving an animal an indelible identification mark or number (fire branding, freeze branding or chemical branding).

Broiler: bird bred for meat production.

Broody: the stage in a bird's productive life when it has finished laying and wants to sit on (incubate) eggs.

Browsing: picking mouthfuls of feed from here and there as the animal moves. Also the name (browse) given to the shrubby material eaten.

Calcify: to be hardened by deposits of lime salts – to become brittle.

Cannibalism: an animal pecking at or feeding on its own species.

Capon: castrated male chicken.

Carrion: decaying or decomposing matter.

Cast: a dog's action to go out and surround sheep.

Castration: removal of the gonads (testicles or ovaries).

Catheter (for AI): a tubular instrument, for depositing semen in the uterus, for drawing blood from vein and artery.

Cellulose: material which forms the solid framework of plants.

Centroid: the central place in a delineated group or area in which a group is dispersed.

Cervix: the entrance or neck of the uterus.

Clitoral flashing (winking): exposing the clitoris.

Clitoris: female's erectile tissue situated at the front of the vaginal opening.

Cloaca: common canal for secretion of wastes and the deposition of semen in birds and amphibia.

Clutch: a number of eggs laid on consecutive days.

Cock: a mature male chicken.

Coitus: mating. See *Copulation.*

Colic: a condition in the horse where severe spasmodic pain of the abdominal organs is present.

Colon: part of the large intestine where cellulose is digested in the horse.

Colostrum: the first milk produced by a dam after birth. Rich in antibodies.

Copulation: the mating act.

Cord: the umbilicus or umbilical cord.

Coronet: the ring or burr round the base of the antler of deer, it forms anew with antlers each year.

Corticoids: name for group of substances secreted from the adrenal cortex or having an influence on that organ.

Covert: actions not easily seen or openly displayed. Hidden or secretive action.

Creche: a nursery or pool of young animals cared for in a group.

Creep feed: food fed to young animals in such a way that the dam cannot eat it. The young creep under a barrier to reach it.

Crook: a metal or wooden stick with a hook on the end used to catch sheep or lambs either by the hind leg or the neck.

Crush: a small pen in which animals can be held tightly, or a part at the end of a race in which animals are restrained.

Cryptorchid: male animal in which the testes have not descended into the scrotal sac but have remained within the abdomen.

Cull: an unwanted (poor) animal. To remove unwanted animals from a group.

Dagging: removing the wool on which faeces has accumulated at the hind end of the sheep.

Dingo: Australian wild dog.

Dipping: immersing animals in a solution of insecticide to kill ecto-parasites (external parasites).

Diurnal: a daily recurring rhythm of activity.

Docking: removing a part of an animal's tail to stop faeces (dung) sticking to it. The short part of the tail remaining is called the dock.

Dog sitting: sitting like a dog. An abnormal posture for a cow, sheep or pig.

Dominance: where some animals because of age, size or sex can dominate other animals and gain access to limited resources such as the best mates, feeding or watering sites.

Drafting: removing certain animals from a group.

Drenching: administering a liquid (usually a drug) down the throat of an animal. Dosing.

Drift-lambing: where unlambed ewes are walked slowly or drifted out from newly lambed ewes with their lambs, which remain behind.

Driving: the action of moving animals ahead of a dog or stockman.

Droving: see *Driving.* Moving stock along to a determined destination.

Dry: not lactating, e.g. dry sow.

Dystocia: difficulty experienced by the dam in the process of giving birth.

Ectoparasite: a parasite on the outside of the animal. (Endoparasite is on the inside of the animal e.g. in the intestine.)

Ejaculation: expulsion of spermatozoa from the penis.

Elimination: a term to cover both urination (voiding of urine) and defecation (voiding of faeces).

Entire: an uncastrated male.

Epididectomy: removal of the epididymus.

Epididymus: duct which serves as the outlet for the spermatozoa or semen.

Epimeletic: social behaviour which relates to the giving of care and attention to the young or new born.

Escutcheon: the skin at the rear of a cow below the rectum and between the hind legs.

Et-epimeletic: social behaviour shown by the young to elicit care-giving responses from older animals.

Ethogram: an exact catalogue of all behaviour patterns occurring in an animal species.

Ethology: the biology of behaviour. Zoological term for the science of animal and human behaviour.

Eye: the characteristic in a working dog where it concentrates intensely on stock and their movements while it is working.

Farrow: to give birth in a pig.

Feed conversion efficiency: the amount of production from an animal (e.g. kg of weight-gain or dozens of eggs) for each kg of feed eaten.

Feedlot: an area where animals are confined and where feed is brought to them.

Feral: formerly domestic but returned to the wild or semiwild condition.

Flehmen: a mammalian response in which the mouth is opened, the upper lip retracted to expose the teeth while the head is lifted high or jerked backward. Occurs frequently in the reproductive season – also termed the lip-curl.

Foetus: the unborn offspring.

Forced moult: deliberate moulting of birds by drastic changes in food and environment to give the bird's reproductive system a rest before a further period of lay.

Fostering: making a dam accept an offspring other than its own, or giving an offspring to another dam (fostering on).

Freeze: the ability of an animal to stay very still to avoid discovery.

Gelded: to be castrated. A gelding is a castrated male horse.

Genitalia: the reproductive organs.

Gentling: stroking or handling of an animal, usually in early life, in a gentle manner to assist tameness and approachability.

Gilt: female pig up to time of first litter.

Gnotobiotic (germ free): pigs born by extracting them directly from the sow inside the sow's uterus. Not exposed to the farm environment so have minimal diseases.

Gonadotrophin: substances which have a stimulating effect on the gonads *viz.* hormones from the pituitary gland and placenta.

Habituate: to learn passively, to respond less often or not at all to repeated stimuli which do not harm or reward the animal, i.e. biologically unmeaningful events.

Haivers: male deer which having been castrated during the first year of life, do not grow antlers.

Harem: a group of females gathered around a male, which then prevents other males from mating with them.

Harnessed teaser: a teaser fitted with a harness so that it marks any female which it mounts.

Head: the action of a working dog to go around and gather stock or stop them straying.

Heat: see *Oestrus*.

Heel: the ability of a dog to bite the heels of cattle to move them along.

Heifer: a young female cow between 4 months and 2 years of age.

Hierarchy: an ordering of group members in relation to social activities like movement (leader–follower hierarchies), or social status (social dominance hierarchies).

Hogget: sheep approximately 6–12 months old.

Hormone: secretions from special ductless glands direct into the bloodstream within the animal that affect various functions.

Hummel: a deer without horns or antlers. Polled but fertile male.

Huntaway: special type of New Zealand dog used for mustering or driving stock. They are trained to bark on command.

Hypocalcaemia: low blood calcium causing milk fever.

Hysterectomy: removal of the uterus.

Imprinting: very rapid learning taking place during a sensitive period – usually near birth or close after puberty which bonds an animal for life to its own species or a substitute. Responses in later life are directed towards the imprinted object.

Ingestive: behaviour which involves the taking into the body of food, water, or body-building substances.

Inguinal: related to the area of the groin.

Inguinal thrusting: thrusting towards/from the area of the groin.

Intromission: the act of placing the male penis inside the female vagina.

Involuntary: action uncontrolled by the mind or brain cortex – ongoing like the heart beating.

Join: the act of putting males and females together for mating.

Judas sheep: a sheep used to lead others.

'King' bull: the most dominant male in a herd which is able to fight off intruders and do most of the mating.

Latency period: the resting period after the male has ejaculated.

Leading dog: dog to go in front of a mob or flock to control its speed and stop animals in it breaking or stampeding.

Let down: the release of milk into the udder system.

Libido: sexual drive or urge to mate. Usually refers to male in farm animals.

Lift: action of a dog to move sheep from a standing position and bring them to the handler.

Lignin: a mixture of substances deposited to form thickened cell walls in woody plants.

Loafing: standing still and appearing to be doing nothing.

Malpresentation: a presentation of an unborn animal in the birth canal where it is wrongly directed. It may have a leg or head back instead of the normal head and front legs first.

Mate: the action of a male mating or serving a female.

Matriarchal dominance order: the rank order of animals based on their ability to pursue or maintain a dominant position in family groups when adult males are not present but adult females are.

Meconium: the faeces produced by the young unborn animal while inside its dam's uterus. Foetal faeces.

Metronome: instrument which beats out time by means of a swinging pendulum.

Monocaste group: socially equal animals, e.g. castrates, or of the same sex all held in the one herd or flock.

Mother-up: action of making sure that each dam has an offspring and that they are correct.

Moult: losing or shedding feathers or hair annually.

Mulesing: removal of wrinkled skin from around the crutch of a Merino sheep.

Mustering: gathering or collecting stock into a group.

Mutilation: to have an organ, i.e. tail or testicles cut off from the body.

Necrosis: death of certain areas of body tissue within a living animal.

Neonate: the newborn at or near birth.

Nocturnal: animal that is active at night.

Nose: a dog's action to bite animals on the nose to turn them.

Noxious: animals that are unwanted pests which damage the environment.

Nymphomaniac: female animal in continuous oestrus. In cattle called a 'buller'.

Oestrogen: generic term for female sex hormones which induce oestrus.

Oestrogenic: having the effect of oestrogen.

Oestrus (the noun): the time of heat in the female when she will stand to be mated. *Oestrous* (the adjective): from oestrus.

Olfactory lobe: the part of the brain concerned with smelling.

Omnivorous: an animal that eats both animal and plant material.

Organic matter: substances which form compounds which are the basis of living matter.

Ovary: the female organ that produces ova (eggs).

Overt: as displayed in the open for all to see.

Oviduct: the receptacle below the ovary in which eggs are dropped before travelling down to the uterus.

Ovulation: the act of shedding eggs from the ovary.

Paddock: grass area fenced off for animals to graze – a field.

Parlour: a building in which cows or goats are milked. There are many designs.
- rotary: where animals stand in a circle (heads in or heads out) and rotate past the milker.
- herringbone: where animals stand side by side at an angle to the milker.

Partogram: a graphic recording of the process of labour proceeding till the point of birth.

Parturition: the act of birth.

Pause: rest period in a bird's laying cycle.

Peck order: order of social dominance in birds or animals.

Pedicle: a permanent bony outgrowth from the frontal bone of the skull from which the antlers grow.

Pedometer: an instrument to record distance travelled by counting steps taken.

Pendulous: hanging down. An udder with poor suspensory ligaments which droops.

Perianal: the area around the anus.

Perineum: region of the body between the anus and the vulva or scrotum.

Peripheral vision: the animal's ability to see at the sides of its eye.

Pheromone: a hormone-like substance produced in distinct glands and released outside the body to transmit information between individuals of the same species – also called social hormones.

Piosphere: an ecological system established in drier rangelands which focusses about a watering point from which foraging animals move, and return usually once or twice each day.

Placenta: the organ inside the uterus by which the dam's blood supply reaches the foetus through the umbilical cord.

Placental membranes: (as above).

Plasma corticoids: corticoids circulating in blood plasma; as distinct from urinary corticoids.

Polydipsia: excessive thirst sometimes resulting from restrictions or stress on a bird or animal.

Precocial: born in an advanced state of development, *viz.* being able to stand up, move and accompany the dam within minutes or hours of birth.

Predator: one which pursues an animal to capture it as prey.

Prehension: the ability to grasp herbage using the tongue and teeth to bite it off.

Pre-orbital gland: a gland located near the front of the eye.

Prepuce: the sheath in which the penis is protected. It has a diverticulum or bag-like structure in which the penis is held.

Progeny test: evaluation of an animal by examining the performance of its progeny.

Promiscuous: an animal that will mate with many other different animals.

Pro-oestrus: the period immediate before oestrus. See *Oestrus.*

Prostatic: related to glands which are accessory to the male sexual organs.

Pseudo pregnancy: a false pregnancy. Acting as if pregnant, when not.

Puberty: the stage of onset of sexual maturity.

Race: a long narrow pen in which animals are held or are moved. A fenced roadway through a farm. Chute (USA).

Random: in no set order. Dictated solely by chance.

Regurgitate: to bring up food again from the stomach or the rumen.

Ruminant: a cud-chewing animal with a rumen or first stomach for fermentation. *Non-ruminant:* animals without a rumen.

Rumination: the act of re-chewing swallowed plant material after regurgitation.

Runt: a smaller, poor growing animal in a litter.

Rut: the period in deer and goats when the males are sexually active and mate the females.

Scrotum: the bag in which the testicles are held.

Season: heat or oestrus. See *Oestrus.*

Semen: the male sperm and fluids produced in the testicles and other glands of the male's reproductive tract.

Shedding: the act of antlers dropping off a stag. The act of a working dog to separate some animals from the group.

Social facilitation: behaviour whereby others are enabled to engage in similar action more easily. Any influence exerted on another which increases the frequency of current responding.

Socialisation: to develop social attachments to species members during a sensitive period in young life – less rapid than imprinting – occurs in altricial or slower-developing species rather immature at birth.

Species-specific: unique species attributes, *viz.* behaviour patterns found in the majority of the species members when in similar places under similar conditions.

Sperm: the male reproductive cells.

Spermatic cord: the tube between testicle where sperm are stored and the penis.

Steer: a castrated male bovine (cattle beast).

Sternum: related to the breast bone.

Stocking density (stocking rate): the number of animals on a particular defined area.

Stud: a term used to describe an animal which is registered with an official breed association. A male horse that does most of the mating in a herd.

Symbiotic: that which mutually benefits two or more partners.

Tail flagging: the act of holding the tail up high in horses.

Teaser: an entire male that cannot ejaculate sperm from its penis. This is usually prevented by surgery.

Tidbiting: pecking at or thrusting the beak towards small or choice materials or food items on the ground.

Traits: characters or characteristics.

Turning tail: where the dog turns away from sheep in a complete circle, rather than facing up to a confrontation.

Umbilical cord: circulatory vessels which link dam and foetus up until birth.

Urethra: the tube joining the bladder and the penis or vagina down which urine flows.

Urophygial: the gland in birds on the tail that produces oil used in preening and waterproofing the feathers.

Uterine inertia: a delay or complete stoppage of normal contractions due to some outside stressor on the dam in labour.

Uterus: the organ in the female in which the offspring or foetus develops during pregnancy. Also called the womb.

Vascular tissue: tissue that carries blood vessels.

Vasectomise: the operation to ligature the spermatic cord in the male and hence make it infertile.

Velvet: early vascular growth of antlers before they harden off, characterised by a soft velvet coating which dies and is rubbed off when the antlers are mature (hard antler).

Vibra recorder: an instrument with a pendulum which marks a chart at any time vibrations of the body cause it to move. Used for recording grazing or movement in animals.

Viscosity: the resistance of a liquid to flow. The more viscous a liquid, the slower its rate of flow.

Voluntary: intentional actions under the control of the animal. Contrasts with involuntary which are ongoing actions not controlled by the brain cortex.

Vulva: the outside lips of the female vagina.

Wallow: to dig and roll in holes full of muddy water.

Waltzing: a courtship response of roosters which is dance-like with one wing dropped and which often leads to crouching in the hen and mating.

Weaning: the removal of an offspring from its dam, or the period when liquid feeding (e.g. from a bucket) stops.

Whelp: the act of producing pups.

Yearling: an animal around one year of age.

Author Index

Subject Index